The Solid-State Mindset

Lessons, Surprises, and Strategies from the Front Lines of Drug Development

Víctor M. Díaz Pérez

solitek
Publishing

Copyright © 2026 Víctor M. Díaz Pérez

All rights reserved.

No part of this publication may be reproduced, stored in a retrieval system, or transmitted in any form or by any means—electronic, mechanical, photocopying, recording, or otherwise—without the prior written permission of the author. Brief quotations may be used in critical articles or reviews.

First edition, 2026
Published by Solitek Publishing
'Solitek' is a trademark of Solid Technologies, S.L.

The information in this book is provided for educational and informational purposes only. It does not constitute professional, regulatory, or legal advice. While every effort has been made to ensure accuracy, the author assumes no responsibility for errors or omissions. Readers are responsible for verifying information and applying appropriate professional judgment.

ISBN (Paperback): 978-84-09-80783-3
ISBN (Ebook): 978-84-09-80784-0

Depósito Legal: SE 3640-2025

Cover design by Víctor M. Díaz Pérez

Printed in the European Union
10 9 8 7 6 5 4 3 2 1

Contents

INTRODUCTION 4
PART I – THE SOLUBILITY PROBLEM NO ONE CAN IGNORE 8
Chapter 1 – Why So Many Drugs Fail: The Rise of the Insoluble Pipeline 9
Chapter 2 – The 100 mg Approach: A Reality Check for Development Teams 13
Chapter 3 – Practical Solubility and Dissolution Science 17
Chapter 4 – What Solubility Teaches Us About the Molecule 21
PART II – SOLID-STATE CHEMISTRY: THE FOUNDATION 23
Chapter 5 – Why Solid Forms Matter: The Architecture of a Drug 24
Chapter 6 – Solid-State Characterisation: Seeing the Invisible 28
Chapter 7 – Polymorphism: When One Molecule Becomes Many 32
Chapter 8 – Seeing the Molecule for What It Really Is 37
PART III – SOLID FORM STRATEGIES FOR DEVELOPMENT 39
Chapter 9 – How to Choose Between Polymorphs, Salts, and Cocrystals 40
Chapter 10 – Crystallisation Process Development: Turning Molecules into Materials 44
Chapter 11 – Particle Size Reduction: Engineering Surface Area for Dissolution 49
Chapter 12 – Choosing the Form the Molecule Can Live With 53
PART IV – ENABLING FORMULATIONS FOR PRECLINICAL SUCCESS 55
Chapter 13 – Early Formulation: Giving Molecules a Fighting Chance 56
Chapter 14 – Building Preclinical Dosing Vehicles: Species, Routes, and Real-World Constraints 60
Chapter 15 – From Possibility to Exposure: The Real Purpose of Early Formulation 65
PART V – STRATEGY, IP, AND THE BUSINESS OF SOLID STATE 67
Chapter 16 – The Business Value of Solid-State Decisions 68
Chapter 17 – Intellectual Property in the Solid State: Lessons from Paroxetine, Gatifloxacin, and the Courts 72
Chapter 18 – The Future: AI, Predictions, and the Next Era of Solid-State Science 77
Chapter 19 – When Physical Behaviour Becomes Strategy 82
PART VI – LESSONS FROM REAL-WORLD CASES 84
Chapter 20 – Ritonavir: When a Molecule Rewrites the Rules 85
Chapter 21 – Gatifloxacin: Stability, Solubility, and the Surprises in Between 90
Chapter 22 – Paroxetine: When the Solid State Meets the Law 94
Chapter 23 – What the Molecules Tried to Teach Us 98
PART VII – FUTURE DIRECTIONS AND FINAL THOUGHTS 100
Chapter 24 – When to Seek Expert Help: Timing, Blind Spots, and Avoiding the Avoidable 101
Chapter 25 – The Future of Solid-State Science: What We Still Don't Know, and Why That Matters .. 105
Chapter 26 – A Career in the Solid State: Reflections, Lessons, and the Road Ahead 109
Chapter 27 – Carrying the Mindset Forward 113
Acknowledgements 115
About the Author 117
References 119

INTRODUCTION

There is a moment in almost every drug development project where the science, the optimism, and the investment meet a simple but unforgiving question: will this compound *actually dissolve?*

It is a question most teams ask too late.

I did not set out to write a book. But after years working with discovery chemists, CMC leaders, formulation scientists, and generic manufacturers, I began to notice a pattern. Project after project—regardless of company size, therapeutic area, or level of experience—kept running into the same set of problems. Some arrived quietly, almost politely: a disappointing dissolution curve, an unexpected humidity sensitivity, the suspicion that a polymorph might be lurking somewhere just out of sight. Others announced themselves loudly: a preclinical formulation that failed to deliver exposure, a batch on the verge of collapsing during scale-up, a promising molecule behaving, quite literally, like brick dust.

And in every case, the root cause could be traced back to the same place: a misunderstanding or an underestimation of solid-state science.

It became clear to me that this wasn't simply a technical gap. It was cultural. Companies invest enormous resources into biology, synthetic chemistry, high-throughput screening, computational design, and regulatory strategy. But the physical reality of the molecule—the way it packs in the solid state, how it interacts with humidity, what its true solubility is, how it responds to stress—often receives attention only when something goes wrong. By then, the consequences can be severe: reformulation, delays, programme resets, missed milestones, or, in extreme cases, crises like the ritonavir polymorph that emerged after launch and shook the industry.

After seeing the same obstacles appear repeatedly, I felt an increasingly strong need to bring clarity to this space. Not to add another highly technical textbook, there are excellent ones already, but to write something more practical and more honest. A book that explains *why* things go wrong, *how* they can be anticipated, and *what* we have learned collectively over decades of successes, failures, and surprises. Something that speaks to the people who must make decisions under tight timelines and imperfect information: biotech innovators trying to keep programmes alive, heads of CMC balancing dozens of risks at once, and generic companies searching for new opportunities in an increasingly competitive landscape.

Solid-state chemistry is often perceived as a world of fine details, i.e. crystallographic planes, transition temperatures, subtle shifts in lattice energy. And while those details matter, what matters even more is understanding what they mean for real projects, real formulations, real scale-ups, and real patients. I have watched brilliant teams struggle not because their science was poor, but because they were fighting against the physical properties of their own molecule without fully realising it. I have also seen the opposite: how a well-chosen solid form, identified early and understood properly, can transform a programme, strengthen its value, and remove entire layers of risk.

This book grew out of conversations, hundreds of them. Conversations with clients trying to make sense of contradictory data, with colleagues debating crystallisation strategies, with formulation teams trying to decide whether an amorphous dispersion was a solution or a last resort. It also grew out of frustration. Too often, issues arising at the interface between solid-state and formulation are seen as bad luck, when in reality they are predictable, manageable, and, in many cases, preventable.

We work in a field where the questions are deceptively simple, yet the answers require a blend of experience, intuition, and scientific discipline. Why does this API dissolve so slowly? Why did a new polymorph suddenly appear? Why does the compound behave beautifully at gram scale but fall apart at the kilo scale? Why does the preclinical formulation work in rodents but fail in dogs? Why does the generic version choose a different hydrate? And why does the same error keep happening across companies, across drugs, across therapeutic areas?

The more I listened, the more I realised that a great deal of wisdom exists in the solid-state community, but it is scattered, across papers, across disciplines, across companies, and often inside the heads of people who have spent a lifetime working in this field. This book is an attempt to bring that wisdom together in one place. Not as a set of rules, but as a set of insights. Not as a manual, but as a guide.

Above all, it is an invitation to step back and look at the molecule not only as a structure or a pharmacophore, but as a physical object that behaves according to its own rules. When we understand those rules, drug development becomes clearer, faster, and far less painful. When we ignore them, the molecule will remind us, sometimes gently, sometimes forcefully.

If this book helps even a few teams anticipate problems earlier, make more confident decisions, or recover time that would otherwise be lost, then it will have served its purpose. Solid-state chemistry will never eliminate uncertainty entirely. But with the right understanding, it can transform uncertainty from a threat into a navigable landscape.

That is why I wrote this book. And that is where our journey begins.

PART I – THE SOLUBILITY PROBLEM NO ONE CAN IGNORE

Chapter 1 – Why So Many Drugs Fail: The Rise of the Insoluble Pipeline

Every generation of drug discovery faces its own challenges. In the 1980s and 1990s, the struggle was identifying meaningful biological targets. In the early 2000s, the focus shifted to combinatorial chemistry and high-throughput screening. Today, we live in the era of exquisitely potent molecules, complex structures capable of hitting very specific pathways with remarkable precision.

But the more imaginative our medicinal chemistry becomes, the more fragile the physical reality of these molecules tends to be. The industry is now grappling with a paradox: our most sophisticated drug candidates often behave, from a physical perspective, like poorly soluble powders that stubbornly refuse to interact with water. In other words, the pipeline has become increasingly insoluble.

Ask any formulation scientist, and they will tell you the same thing: modern small molecules dissolve slowly, incompletely, or unpredictably. Many barely dissolve at all. And yet these compounds are expected to perform in biological environments far more complex than any laboratory flask. The disconnect between potency on paper and viability in real human physiology has grown wider year after year.

There are many reasons for this. Drug design has moved toward molecules that are heavier, more lipophilic, more rigid, and more aromatic, because those traits often improve potency, binding affinity, or selectivity. Unfortunately, they also tend to reduce aqueous solubility. The physicochemical properties that make a drug a strong binder frequently make it very insoluble. Medicinal chemists know this, of course, but early optimisation is inevitably centred on biological performance. Solubility is often seen as something that can be fixed later.

The problem is that *later* always arrives. Sometimes it arrives quietly, when a compound performs beautifully *in vitro* but fails to produce meaningful exposure *in vivo*. Sometimes it arrives suddenly, when a preclinical study produces inconsistent plasma levels. And sometimes it arrives dramatically, when a formulation that should work, by all conventional logic, simply doesn't.

The shift from the old BCS classification to the refined developability classification system reflects this growing awareness. For decades, Class II compounds were loosely grouped together, united by one trait: low solubility. But as more molecules began to fall into this category, it became clear that solubility and dissolution behave very differently, and that lumping these molecules together was masking critical distinctions. We now understand that some drugs fail because they dissolve too slowly, while others fail because they cannot dissolve beyond a very low thermodynamic threshold. The strategies for addressing each issue are not the same, and misunderstanding which problem you have can cost years.

Solubility problems are not academic curiosities. They shape the entire development journey. They determine whether a compound can be dosed orally, whether a formulation is feasible, whether exposure can be achieved in animals, and whether the drug can move confidently into human trials. They influence manufacturing, stability, dose size, patient experience, and ultimately commercial viability. And, as the ritonavir crisis famously showed, they can even threaten a drug after approval.

Yet, despite all this, solubility is still one of the most underestimated risks in early development. Projects with outstanding biological data can, and frequently do, stall because teams assume that solubility can be *handled* as needed. They trust that particle size reduction, amorphous dispersions, lipid systems, or a clever formulation will save the day. Sometimes they do. Just as often, they do not. The later the issue is discovered, the harder it becomes to solve.

The tragedy is that poor solubility is almost always predictable. Long before a compound reaches IND-enabling studies, its physicochemical profile already tells a story. The challenge is not the lack of information, but the tendency to overlook it. Teams are under pressure to move quickly, to generate data, to keep the programme alive. Early physical characterisation can feel like a luxury, or worse, like a delay. But nature does not negotiate with timelines. If a compound dissolves at five micrograms per millilitre, no amount of optimism will make it soluble at ten.

In recent years, I have seen companies lose enormous value because they reached these realisations too late. I have also seen programmes rescued because a single scientist had the instinct to ask the right question early enough. What separates these outcomes is rarely the sophistication of the technology. It is the timing of the insight.

The insoluble pipeline is not a temporary trend; it is the new reality of small-molecule innovation. It demands a shift in mindset, a recognition that solubility is not simply another box to tick but a defining attribute of a molecule's developability. Understanding it early is not conservative, it is strategic. And ignoring it is not efficient, it is costly.

In the chapters that follow, we will explore why this has happened, how solid-state science offers a powerful framework for navigating the problem, and how early decisions can shape the success or failure of a drug years later.

The story of every molecule begins long before its first clinical dose. And more often than not, that story begins with how it behaves in the solid state.

Chapter 2 — The 100 mg Approach: A Reality Check for Development Teams

Every development programme reaches a moment when enthusiasm meets arithmetic. For many molecules, that moment arrives in the form of a deceptively simple question: *How much of this compound will a human being actually need to take?*

In practice, the numbers that matter are not the elegant ones that appear on pharmacology slides. They are the ones that tell us whether a patient can swallow a reasonable dose, whether a capsule can be manufactured, and whether the drug can achieve therapeutic exposure without relying on wishful thinking. Among all these numbers, one has become particularly symbolic: the 100 mg mark.

The 100 mg "approach" is not a rule, nor a regulatory guideline. It is a reality check, a way to anchor discussions about feasibility. When a molecule requires doses well under 100 mg, many options remain open. Solubility challenges can often be managed; enabling technologies can compensate for limitations; complex formulations may still be practical. But once the projected human dose begins to climb beyond 100 mg, everything becomes more constrained. Formulation options narrow, the margin for error shrinks, and the consequences of poor solubility multiply.

This is where teams often misjudge their situation. A compound that dissolves slowly but is active at a low dose may still be viable. But a compound with marginal solubility that requires hundreds of milligrams to reach therapeutic exposure becomes much harder to formulate, and far more likely to fail. Dose is not simply a number, it is a boundary condition that defines what strategies are still available.

To anyone who has spent time working in early development, the pattern is familiar. A compound shows beautiful potency *in vitro*. It performs well in the first *in vivo* study, often in rodents, where fast transit and high metabolism sometimes mask deeper issues. The project advances quickly; timelines tighten; enthusiasm grows. It is only later, sometimes much later, when more rigorous pharmacokinetic data arrive, that the required exposure becomes clear – and with it, the real dose.

Suddenly, the compound that looked manageable at 10 mg/kg in mice now requires 500 mg per day in humans. And now the team discovers that its solubility sits somewhere between "*very low*" and "*practically non-existent.*" In that moment, the 100 mg threshold stops being an abstract concept and becomes a barrier. The question that nobody wanted to ask early on now becomes unavoidable: *Is this dose actually feasible?*

The 100 mg approach forces this conversation earlier, when there is still time to act. It encourages teams to consider not just what the molecule does biologically, but what it demands physically. If a compound dissolves at only a few micrograms per millilitre, scaling that behaviour into a human gastrointestinal tract requires a sober evaluation of what is realistic and what is not. Even the most sophisticated formulation technologies – amorphous dispersions, nanomilling, lipid systems – have limits. They cannot magically compensate for a compound that fundamentally resists dissolution.

There is another dimension to this problem that is easy to overlook: the variability introduced by physiology itself. A drug that requires very high doses to overcome poor solubility is often more sensitive to food effects, transit time, pH changes, and inter-patient differences. As dose increases, the formulation must work harder to compensate for those variables. When the margin for error disappears, clinical data become unpredictable, and unpredictable drugs become unviable.

The point of the 100 mg approach is not to discourage bold chemistry or complex molecules. It is to bring realism into discussions that too often drift into optimism. Drug development rewards early clarity, not late heroics. The earlier a team understands the physical burden their molecule imposes, the better equipped they are to decide whether to move forward, modify the chemistry, consider alternative solid forms, or invest in enabling strategies before time is lost.

What makes this threshold so powerful is not its exact value, but what it represents. It reminds us that every molecule has physical limits, and that those limits cannot be ignored indefinitely. It challenges teams to confront uncomfortable truths at the only moment when doing so is still inexpensive. And it exposes a simple but often forgotten fact: formulating a drug is much easier when the dose is small.

Some compounds will justify heroic formulation strategies. Others will not. The 100 mg approach helps distinguish between them long before the hard decisions become irreversible.

In the end, it is not a test the molecule must pass. It is a conversation the development team must have early, honestly, and with a clear understanding of what is feasible. When that conversation happens in time, programmes become more predictable, timelines shorten, and fewer surprises appear when the stakes are highest.

The story of solubility is always, in some sense, a story about dose. And the sooner we acknowledge that relationship, the fewer promising molecules we will lose to problems that were visible from the very beginning.

Chapter 3 — Practical Solubility and Dissolution Science

Solubility is a word everyone in drug development uses, but not always in the same way. It is a deceptively simple concept: how much of a substance dissolves in a given medium. Yet behind that simplicity lies one of the most persistent sources of confusion in early development.

A compound can be highly potent, structurally elegant, and synthetically accessible. But if it refuses to dissolve in the gastrointestinal tract, all those strengths become irrelevant. The relationship between a drug and water governs everything from its dose to its formulation to its clinical behaviour. And still, solubility is often discussed as though it were a fixed number that can be measured once and filed away. In reality, solubility is a landscape, one shaped by pH, ionisation, temperature, solid form, particle size, and the dynamic conditions of the body itself.

One of the first misunderstandings teams encounter is the difference between solubility and dissolution. Solubility tells you the maximum amount of drug that can be held in solution at equilibrium. Dissolution tells you how fast the solid dissolves as it approaches that limit. These two numbers behave differently, are influenced by different factors, and dictate very different formulation strategies. Yet it is surprisingly common for teams to treat them as interchangeable. When early data show poor oral exposure, some assume the drug is inherently insoluble, when the real problem may lie in how quickly it dissolves. Others believe dissolution can be "*fixed*" with particle size reduction, when the true limitation is that the compound cannot exceed a very low thermodynamic ceiling, no matter how small the particles become.

This distinction became much clearer with the refined developability classification system, which divides the old BCS Class II category into two types. Class IIa compounds are limited primarily by their dissolution rate. Class IIb compounds are limited by their solubility itself. The difference may seem subtle, but the consequences for development are significant. A Class IIa drug might be saved by reducing particle size or improving wettability. A Class IIb drug usually requires a more transformative approach, such as a salt, a cocrystal, or an amorphous dispersion. Putting a Class IIb compound through a dissolution-focused strategy is like trying to fix structural cracks in a building by repainting the walls. It may look better temporarily, but the underlying problem remains.

The behaviour of a compound during dissolution is further complicated by the conditions of the gastrointestinal tract. Most small molecules experience different pH environments as they travel from stomach to intestine. Weak bases often dissolve readily in the acidic stomach, only to precipitate in the higher pH of the intestine, where absorption actually takes place. Weak acids sometimes show the opposite pattern. A single solubility value cannot capture this dynamic. Nor can it predict supersaturation, a temporary state in which more drug is in solution than thermodynamics would suggest, only for it to crash out moments later. Some compounds benefit from supersaturation; others suffer from it. Many producing promising dissolution profiles in vitro fail to maintain those conditions long enough in vivo to deliver meaningful exposure.

Then, there is the question of the solid form. A polymorph, a salt, or a cocrystal can each produce a different solubility profile, not only in terms of magnitude but in how the concentration evolves over time. Some forms dissolve quickly but fall out of solution just as fast. Others dissolve slowly but remain in solution longer. Some change during dissolution, converting into a more stable but less soluble form. The behaviour of the solid phase is not disconnected from the dissolution profile; it shapes it continuously.

The industry often talks about solubility as though it were a problem belonging solely to formulators. But long before a formulator touches the compound, its solubility has already written the first draft of its development story. This is why the earliest solubility measurements, however rudimentary, are so important. They reveal the width of the path available. They hint at whether the molecule will behave predictably, or whether it will require careful engineering to coax it into a workable formulation. They tell us whether a 50 mg dose is plausible, or whether the compound is heading toward territory where delivery becomes impractically difficult.

What makes solubility particularly challenging is that it does not exist in isolation. It interacts with permeability, metabolism, stability, transit time, and even patient variability. A compound with marginal solubility can sometimes succeed if everything else aligns perfectly and absorption is rapid. But few molecules enjoy that kind of cooperation. More often, solubility becomes the bottleneck that controls the entire therapeutic experience, from the amount absorbed to how consistently patients respond.

The most successful development teams are not the ones with the most sophisticated technologies, but the ones that learn to read these early signs for what they truly mean. They do not take a dissolution curve at face value; they look for the hidden story behind it. They ask whether the compound is limited by thermodynamics or by kinetics. They examine pH-dependent behaviour not as a box to tick, but as a roadmap that predicts how the molecule will behave in the body. And they do all of this before committing to a formulation strategy that may be misaligned with the molecule's actual needs.

Solubility and dissolution are often presented as obstacles. In reality, they are sources of information. They reveal the strengths and weaknesses of the molecule. They shape the strategy rather than frustrate it. And when understood properly, they give teams the rare gift of foresight, the ability to anticipate challenges long before they become expensive or irreversible.

In the next chapters, we will move beyond these concepts and explore the solid forms themselves, the architecture that defines how a molecule behaves in the real world. Because while solubility sets the stage, it is the solid state that determines how the story unfolds.

Chapter 4 — What Solubility Teaches Us About the Molecule

Solubility is often described as a property, a number, a constraint... but in truth, it is a conversation. It is the first moment where the molecule begins to reveal what kind of partner it will be in development. After years watching compounds struggle, behave unpredictably, or succeed beyond expectation, the pattern becomes unmistakable: solubility is rarely the problem teams think they have, and almost always the one they discover too late.

The chapters in this part were meant to shift perspective. To remind us that the insoluble pipeline is not a passing trend; it is the landscape we must navigate. That dose is not a downstream detail but a defining condition. And that dissolution is not a formality but a physical negotiation between solid and solvent, a negotiation the molecule will not always win.

If there is a call to action here, it is simply this: listen early. Before models, before formulation brainstorming, before rescue strategies, listen to the numbers the molecule is already giving you. They are rarely misleading. Poor solubility, slow dissolution, pH-dependent behaviour... each one is a signal, a hint at the physics ahead.

Early clarity does not eliminate difficulty, but it compresses uncertainty into the moment where it can still be shaped. And in the long, often unpredictable world of small-molecule development, that is one of the most valuable advantages a team can have.

PART II — SOLID-STATE CHEMISTRY: THE FOUNDATION

Chapter 5 — Why Solid Forms Matter: The Architecture of a Drug

Every drug begins its life as a molecule, but every medicine begins its life as a solid. This simple shift from molecular structure to solid-state structure is one of the most important transitions in pharmaceutical development. It is also one of the least appreciated.

Medicinal chemists spend months refining a compound's shape, polarity, flexibility, and binding affinity. They design it to fit beautifully into its biological target, to evade metabolism long enough to do its job, and to avoid off-target effects. But once that molecule leaves the realm of theory and enters the physical world, it must obey a different set of rules. It must decide how to arrange itself in three-dimensional space, how to interact with water, how to respond to temperature and humidity, and how to exist alongside impurities, solvents, and excipients. These behaviours, quiet, microscopic, and invisible to the naked eye, shape everything that follows.

A molecule can crystallise in different ways. It can arrange itself into tightly packed lattices or looser, more open structures. It can incorporate solvent molecules into its architecture, forming solvates or hydrates. It can partner with other molecules to form salts or cocrystals. It can even give up long-range order entirely, existing in an amorphous, higher-energy state. Each of these manifestations is chemically identical, yet physically distinct. And those differences are not subtle; they determine solubility, compressibility, stability, manufacturability, and ultimately bioavailability.

There is something almost architectural about this process. The molecule is the brick, but the solid form is the building. Every crystalline arrangement is a structure with different rooms, corridors, and supporting beams. Some forms are airy and porous; others are dense and rigid. Some welcome water, others repel it. Some collapse at the first hint of humidity, others stand firm under stress. In some, impurities hide easily; in others, they are pushed out. What we call "polymorphs" are simply different ways the same bricks can be arranged into a habitable structure or into one that collapses under its own weight.

When a molecule first crystallises in a laboratory, the form it adopts is often the one that is easiest to produce under those conditions, not the one that is most stable or most developable. Discovering the full landscape of possible arrangements requires careful exploration. Sometimes the first form is ideal. More often, it is not. There are cases where a metastable form dissolves far more readily than the most stable one, making it attractive for early exposure studies. There are cases where a salt unlocks solubility that the free base could never achieve. And there are cases where a cocrystal delivers both performance and stability that neither component can offer alone.

Choosing the right solid form is therefore not an academic exercise. It is a strategic one. It determines what options are available downstream. A poorly soluble form may force the development team into complex formulation technologies that slow timelines and increase cost. A highly hygroscopic form may require special packaging or strict humidity controls. A metastable form may threaten to convert into something very different during processing. Each decision made at this stage shapes the entire development path: the manufacturing process, the dosage form, the supply chain, even the intellectual property position.

One of the most striking lessons from working in this field is how often troubles emerge because a team did not realise that their solid form was changing or could change. A compound may behave flawlessly during early experiments, only to crystallise differently when scaled from grams to kilograms. A seemingly robust polymorph may reveal a transition temperature that makes it unstable under process conditions. A neat, well-behaved salt may disproportionate in the presence of moisture or excipients. These are not unusual events, they are the natural behaviour of molecules navigating their energetic landscape. But without the right understanding, they appear unpredictable, even mysterious.

Amorphous materials add another dimension. They offer the promise of higher apparent solubility and faster dissolution, often breathing life into molecules that would otherwise fail. But they also carry inherent instability. To exist in an amorphous state is to exist in a high-energy, restless configuration. Given time, temperature, or humidity, these systems tend to find their way back toward a more stable crystalline arrangement. Stabilising them requires polymers, careful formulation, and an acceptance that the rules governing them are different from those governing crystals. When used well, they can transform a programme. When used without understanding, they can create more problems than they solve.

This interplay between solid form, solubility, and stability is not static. It evolves throughout development. Early materials, often produced under medicinal chemistry conditions, may bear little resemblance to the materials prepared during process optimisation. A form that appears stable on the bench may behave differently in the reactor. What dissolves easily at milligram scale may not dissolve at all at manufacturing volumes. The solid-state landscape is dynamic, and development teams must remain alert to its shifts.

Understanding why solid forms matter is ultimately about acknowledging that molecules behave differently once they crystallise. They reveal preferences, vulnerabilities, and strengths that are not visible on a structural diagram. They behave like materials, not abstractions. And the sooner a team embraces this perspective, the more control they gain over the future of their molecule.

In the chapters ahead, we will uncover the techniques used to explore this landscape, how we measure these properties, how we interpret them, and how they guide the decisions that determine whether a compound becomes a drug or remains an unrealised idea. Solid-state chemistry is often described as complex, but in truth it is intuitive. It is the study of how molecules choose to exist in the real world. And it is hard to imagine a more important question in drug development than that.

Chapter 6 — Solid-State Characterisation: Seeing the Invisible

One of the quiet challenges in solid-state science is that so much of what matters cannot be seen. A tablet looks like a tablet. A powder looks like a powder. To the naked eye, two crystalline forms of a compound appear identical. Yet these invisible differences, the spacing of molecules, the way hydrogen bonds interlock, the presence or absence of water in the lattice, etc. shape the drug's entire behaviour.

Solid-state characterisation is the discipline that lets us see what the eye cannot. It provides the fingerprints of a material, the unmistakable signs that reveal its identity, its purity, its weaknesses, and its potential. When done well, it replaces uncertainty with understanding. When done poorly or not at all, it leaves development teams navigating in the dark.

The first thing most scientists learn when entering this field is that no single technique tells the full story. Each method captures a different dimension of the solid-state landscape. Some probe structure, others probe energy, others probe mass or moisture.

None are sufficient on their own, but together they create a coherent picture, much like different medical imaging techniques combine to diagnose a patient. In that sense, solid-state characterisation is less of a test and more of an investigation.

X-ray powder diffraction (XRPD) is often the starting point. It is the closest one can get to "*seeing*" the arrangement of molecules in a powder. The patterns it produces, i.e. sharp peaks for crystalline materials, broad halos for amorphous ones, are as distinctive as a fingerprint. Two polymorphs of the same molecule will produce different XRPD patterns, even if their chemical structures are identical. In development, XRPD becomes the anchor method, the evidence that confirms whether a solid form is consistent from batch to batch, whether an unexpected phase has appeared, or whether a formulation has remained physically stable. It is one of the few techniques that can answer the most fundamental question: what form do we actually have?

But knowing the form is not enough. A material's behaviour under heat often reveals just as much. Differential scanning calorimetry (DSC) captures the energetic events that occur as a material is heated: melting, crystallisation, glass transitions, solid–solid transformations. These events often tell us not only which form we have, but how it behaves, how stable it is, and how close it is to changing into something else. A sharp melting point can signify purity and well-ordered packing. A broad endotherm may signal a transformation taking place under the surface. A glass transition marks the restless mobility of an amorphous system. Thermal analysis becomes a kind of conversation with the molecule, revealing how it responds to stress.

Thermogravimetric analysis (TGA) adds another layer. Many solids carry water or solvent within their crystal structures, sometimes tightly bound, sometimes held only gently. TGA measures mass loss as a function of temperature, showing exactly when and how much of this volatile content is released. A hydrate may lose water at a predictable temperature; a solvate may release solvent in stages. These behaviours matter because they affect stability, processing, and safety. A form that loses water too easily may transform in humid conditions. A solvate that holds on too tightly may complicate drying or raise regulatory questions. TGA reveals what the eye cannot: the hidden companions inside the lattice.

Moisture itself is its own story. Some molecules embrace water; others resist it. Hygroscopicity can alter a compound's solid form, its dissolution behaviour, its flow, and its compressibility. Understanding how a material interacts with humidity is essential for predicting real-world behaviour. Dynamic vapour sorption studies, storage under controlled conditions, and humidity-stressed experiments all serve the same purpose: to learn how the material behaves when exposed to the air it will inevitably encounter. Sometimes the results are benign. Sometimes they reveal transitions that threaten stability if not carefully managed.

Behind all of these methods lies a simple truth: molecules are not static objects. They respond to their environment, sometimes subtly, sometimes dramatically. Characterisation is how we catch those shifts before they become problems. It tells us whether a material remains stable across temperature cycles, whether grinding induces amorphisation, whether a salt disproportionates under humidity, or whether a cocrystal survives formulation. It turns assumptions into facts.

Perhaps the most misunderstood aspect of solid-state characterisation is that it is not merely confirmatory. It is diagnostic. It reveals why something is happening, not just whether it has happened. If a batch dissolves more slowly than expected, characterisation may show that the form has changed.

If a formulation behaves unpredictably, characterisation may reveal the presence of amorphous material, a small amount of which can influence the whole system. If process scale-up produces inconsistent results, characterisation can identify shifts in morphology or transitions induced by stress. In each case, these tools do not just document the problem, they explain it.

There is a unique satisfaction in seeing a development team gain clarity through characterisation. Confusion gives way to understanding, speculation becomes unnecessary. The solid form that once seemed unpredictable suddenly appears logical, governed by rules that can be managed with the right knowledge. Problems that seemed mysterious resolve into patterns. And with that clarity comes better decisions about form selection, about formulation strategy, about process design.

Solid-state characterisation is sometimes treated as an obligation, a regulatory checkbox. In truth, it is the foundation upon which predictable development is built. It gives us the ability to see the invisible architecture of a drug, to understand its behaviour at a level that chemistry alone cannot provide. Without it, we are guessing. With it, we are navigating with a map.

In the next chapter, we will explore how this map becomes indispensable when dealing with polymorphism, the phenomenon that has humbled even the most experienced teams and reshaped the industry's understanding of risk.

Chapter 7 — Polymorphism: When One Molecule Becomes Many

There is a moment in solid-state science that feels almost like a magic trick: when you realise that a single molecule, unchanged in its chemistry, can nonetheless exist as many different solids, each with its own character. This is polymorphism, the ability of a compound to arrange itself into more than one crystalline form. And once you learn to see it, you understand why it has fascinated, frustrated, and humbled scientists for decades.

To someone outside the field, polymorphism sounds esoteric, almost academic. But within pharmaceutical development, its consequences are anything but theoretical. Polymorphs differ in density, melting point, solubility, stability, hygroscopicity, flowability, compressibility, filterability... properties that influence everything from a drug's manufacturability to its bioavailability. Two forms that look identical to the naked eye may behave as though they belonged to entirely different molecules.

The most enduring lesson about polymorphism is that molecules tend to explore more structural possibilities than we expect. A medicinal chemist might crystallise a compound during discovery and believe, with some justification, that the form obtained is "*the form*". Early results reinforce this comfort: batches behave consistently, initial measurements are reproducible, and nothing suggests a hidden landscape. But beneath that stability lies a diversity of arrangements the molecule could adopt under slightly different conditions, a different solvent, a different cooling rate, a different level of supersaturation, a different seed, a different stress.

The stability of a polymorph is often misunderstood. It is tempting to think of one form as inherently the most stable and the others as fleeting curiosities. In reality, stability exists on a spectrum shaped by temperature, pressure, humidity, and the presence of other molecules. What appears stable at room temperature may be less stable at elevated temperature. Some pairs of polymorphs are monotropic, meaning one form is always more stable than the other; others are enantiotropic, meaning their stability reverses at a particular transition temperature. This complexity is not an exception but a rule.

Polymorphism would be a manageable curiosity if molecules stayed where they were put. But they do not. Under stress, i.e. mechanical grinding, heating, cooling, drying, milling, granulation, humidity exposure, or even simple time, a metastable form may transform into a more stable one. Sometimes the transformation is slow. Sometimes it is fast. Sometimes it is triggered by something as trivial as the presence of a few seed crystals, invisible to the eye but powerful enough to redirect the entire crystallisation pathway.

The appearance of a new polymorph can be dramatic. The most famous case, ritonavir, still sends a shiver through development teams. A more stable crystalline form, unseen during development, appeared suddenly during commercial manufacture and rendered the existing formulation nearly useless. It was not carelessness, it was a reminder that molecules explore possibilities even when we think the landscape is known. The industry learned from that moment, strengthened screening strategies, and refined its understanding of risk. But the underlying truth remains unchanged: polymorphism can be unpredictable, and we ignore that unpredictability at our peril.

Even when polymorphs behave peacefully, their differences matter. A more stable but less soluble form may have slower dissolution, forcing formulators into complex enabling technologies. A metastable but more soluble form may deliver excellent bioavailability but require careful control to prevent conversion. A hydrate may dissolve rapidly but transform in humid air. A solvate may crystallise beautifully but resist drying. Each form is a different balance of energy and constraint, and development teams must decide which balance aligns with the needs of the project.

One of the most subtle aspects of polymorphism is that it often reveals itself only when conditions scale. A crystallisation process that works beautifully in the fume hood may behave entirely differently in a reactor. The rate of cooling may shift. Mixing patterns may change. Supersaturation may build more slowly or more rapidly. What once favoured one polymorph may now favour another. These transitions can feel like surprises, but they are simply the molecule expressing its preferences under new circumstances.

Understanding polymorphism is not about memorising definitions or classifying forms. It is about listening carefully to what the molecule is telling us. If a new form appears during a routine experiment, the molecule is revealing part of its landscape. If a process is inconsistent, the molecule may be transitioning between states. If a formulation behaves unpredictably, the solid form may be shifting under the surface. Polymorphism is not a nuisance, it is a form of communication.

What makes polymorphism both beautiful and challenging is that it reflects the essence of molecules as dynamic, responsive systems. They are not static crystals locked forever in place. They are arrangements of atoms exploring the lowest-energy structure available to them. Sometimes they find it immediately. Sometimes they find it later. Sometimes they never find it unless we create the right conditions. The role of solid-state scientists is not to force the molecule into a form, but to understand its landscape well enough to guide it.

In development, that understanding becomes a foundation for everything that follows. Once the polymorphic landscape is mapped through screening, characterisation, modelling, and experience, the team can choose a form with confidence. They can predict its behaviour, control its production, and design processes that maintain its integrity. They can navigate scale-up without fear of unexpected transitions. And importantly, they can plan the formulation strategy around a solid form that supports the therapeutic goals of the drug.

Polymorphism teaches humility. It reminds us that even the simplest-looking molecules contain a universe of possibilities. But it also offers a powerful opportunity. The ability to choose the right solid form, to control it, and to understand its behaviour is one of the greatest levers we have in drug development. When embraced fully, it transforms risk into strategy and uncertainty into knowledge.

In the next chapter, we will move from this landscape of possibilities to the practical question that every development team must eventually answer: how to choose between polymorphs, salts, and cocrystals, the different architectures a molecule can adopt on its way to becoming a drug.

Chapter 8 — Seeing the Molecule for What It Really Is

Solid-state chemistry is sometimes presented as a set of analytical techniques, a list of instruments, or a catalogue of definitions. But its true value lies in something simpler: it reveals the physical identity of the molecule. XRPD, DSC, TGA, moisture studies... they are not tests to be passed, but windows into how the molecule behaves when the world stops being theoretical.

If solubility is the first conversation with the molecule, the solid state is the first moment you truly see it.

This part aimed to show that the physical arrangement of atoms in the solid form is not a detail; it is the architecture that defines everything that follows. The molecule may look elegant on a drawing, but its crystal structure, its transitions, its hydration tendencies, its energetic preferences are what determine whether development will unfold smoothly or strain under hidden pressures.

The call to action is one of mindset. Approach the molecule as a material, not just an idea. When a team embraces that shift early, analytical work becomes less about confirming assumptions and more about discovering the truth. And truth, even when inconvenient, always makes development easier than uncertainty.

PART III – SOLID FORM STRATEGIES FOR DEVELOPMENT

Chapter 9 — How to Choose Between Polymorphs, Salts, and Cocrystals

At some point in every drug development programme, the question arises: *what is the right solid form for this molecule?* It sounds straightforward, almost administrative. But the truth is that this decision sits at the intersection of chemistry, physics, formulation, manufacturability, and strategy. Choosing the solid form is not simply about finding what works today, it is about choosing the form the molecule can live with for the rest of its life.

Every molecule has an inherent set of possibilities. It can crystallise into different polymorphs, each with its own energy landscape. If it carries ionisable groups, it may form salts with a variety of counterions. If it can form hydrogen bonds or other non-covalent interactions, it may generate cocrystals with suitable partners. These architectures differ not only in how the molecules arrange themselves, but in the properties that emerge from those arrangements. And those properties govern the entire downstream journey.

The simplest path is often to begin with what the molecule does naturally. During early discovery, medicinal chemists may produce a crystalline form without much intention, simply as a consequence of isolating the compound from solution. This initial form is sometimes entirely acceptable: sufficiently stable, reasonably soluble, and easy to handle. But more often, this early form is only a hint of what the molecule could become. It may be a metastable polymorph that dissolves well but transforms easily. It may be a stable form with poor solubility. It may be a hydrate that behaves beautifully in the lab but unpredictably in humid conditions. Treating this first form as definitive is a common mistake.

Salts offer a different route. When a molecule can be protonated or deprotonated, forming a salt can dramatically change its solubility, stability, and even its manufacturability. Many of the most successful oral drugs on the market are salts for precisely this reason. By pairing the molecule with a counterion, one can unlock new crystal structures with different dissolution rates, reduced hygroscopicity, or improved mechanical properties. But salts are not a cure-all. A salt must first be chemically justified, this is, the ΔpKa must allow it, but even when the chemistry cooperates, the resulting salt may be hygroscopic, unstable, or prone to disproportionation. A salt that looks promising on the bench may break down in the presence of excipients or moisture, reverting to the free base or free acid. Salt formation is powerful, but it must be guided, not assumed.

Cocrystals add yet another dimension. They are not salts, because no proton is transferred. Instead, the drug molecule forms an ordered structure with a neutral partner molecule, a coformer, with which it interacts through hydrogen bonds, π - π stack interactions, or other non-covalent forces. Cocrystals can enhance solubility without altering the drug's ionisation state. They can improve stability, modify mechanical properties, or enable processing routes that would be impossible with the pure API. The regulatory environment has also evolved, recognising cocrystals as legitimate pharmaceutical forms rather than exotic curiosities. But cocrystals require insight and intent. They depend on the careful selection of a coformer, and their performance depends on maintaining that supramolecular structure throughout development and storage.

The choice between polymorph, salt, and cocrystal is therefore not a linear decision, but an evaluation of trade-offs. A stable polymorph may offer excellent manufacturability but poor solubility. A metastable polymorph may offer superior dissolution but require stringent controls. A salt may deliver the solubility needed for early studies but introduce new risks around humidity or pH. A cocrystal may provide the ideal balance, but only if its behaviour is fully understood and its preparation is reproducible at scale. No form is inherently superior, each one of them is a tool whose value depends on the context.

In practice, these decisions often unfold under time pressure. A preclinical study is approaching, and a solid form must be chosen quickly. Exposure needs are urgent, and the formulation team must decide whether a salt screen, a cocrystal exploration, or an amorphous strategy will buy enough solubility to move forward. Later, as the molecule advances, the priorities shift: stability becomes critical, manufacturability becomes dominant, and intellectual property considerations emerge. A form chosen for early studies may not be the form that enters the clinic. A form chosen for the clinic may not be the one used in commercial manufacture. Understanding this evolution is essential.

What makes this process both challenging and rewarding is that molecules do not always respond in predictable ways. A salt that appears promising may crystallise in an unexpected polymorph during scale-up. A cocrystal with an elegant hydrogen-bonding network may absorb moisture and transform. A stable polymorph may unexpectedly transition when exposed to mechanical stress. These unpredicted behaviours are not failures but the natural expression of the molecule's preferences. The goal is not to eliminate surprises but to understand the landscape well enough that the surprises can be managed.

One of the most important insights in solid-state science is that the "*best*" form is not the one with the highest solubility or the greatest stability, but the one that offers the right balance for the intended use. A metastable form may be ideal for early toxicology studies, enabling exposure without requiring a complex formulation. A salt may be ideal for clinical supply, striking the right balance between solubility and manufacturability. A stable polymorph may be ideal for commercial manufacturing, ensuring robustness at scale. A cocrystal may be the key to a differentiable product in a competitive generic landscape. The right form depends on where the molecule is going, not just where it is today.

Choosing a solid form is ultimately about listening. Which arrangement does the molecule favour? Which forms are energetically accessible? How does it respond to humidity, stress, or solvents? How does its structure influence its solubility, its dissolution behaviour, its mechanical properties? The best choices come from understanding these answers, not forcing the molecule into a predetermined strategy.

The next chapters will explore how crystallisation brings these decisions to life, how supersaturation, nucleation, growth, and process design shape the forms that appear, and how careful engineering can turn a fragile laboratory observation into a robust, reproducible solid form suitable for development at any scale.

Chapter 10 — Crystallisation Process Development: Turning Molecules into Materials

For a molecule to become a medicine, it must first become a material. This transformation is neither automatic nor trivial. It is mediated by crystallisation, the quiet, intricate process through which molecules leave solution and assemble themselves into the solid forms we depend on. Crystallisation is where chemistry meets engineering, where theory meets reality, and where many of the most persistent challenges in drug development first emerge.

It is easy to underestimate crystallisation. The word evokes images of flasks sitting undisturbed on a bench, crystals slowly appearing as solvent evaporates or temperature drops. In the laboratory, this simplicity is part of its charm. But beneath that calm surface lies a complex interplay of thermodynamics and kinetics, one that determines not only which solid form appears but how that material behaves when the process is scaled from milligrams to tonnes.

At the heart of crystallisation lies a single concept: supersaturation. Molecules only crystallise when the solution contains more of them than it can comfortably hold at equilibrium. This excess creates the driving force for crystallisation, but the system must decide how to release it. It can do so by forming many tiny crystals, this is, nucleation, or by growing the surfaces of crystals that already exist. The balance between these pathways shapes everything: particle size, purity, morphology, filterability, flow, and even the likelihood of forming a different polymorph.

Supersaturation is not a static quantity. It evolves continuously as the solution cools, as antisolvent is added, as crystals grow, as the solute dissolves or precipitates. The same process that produces a clean, well-defined slurry at small scale may produce uncontrolled nucleation or sluggish growth at larger scale simply because the dynamics have shifted. The molecule is the same, but the environment changes, and molecules respond to their environment with remarkable sensitivity.

One of the most telling lessons in crystallisation is that molecules remember their past. A solution that has been heated too quickly, cooled too abruptly, or exposed to seeds of another polymorph may take a completely different path than one handled gently. Small differences in agitation, solvent composition, or even the surface of the vessel can tip the balance towards one form or another. This is why crystallisation processes that seem straightforward in a round-bottom flask can become unpredictable in a reactor. At small scale, a chemist's intuition often compensates for hidden variables. At large scale, everything must be engineered.

Seeding is one of the most powerful tools we have to impose order on this complex landscape. By introducing a small quantity of crystals with the desired form and size, we guide the solution toward growth rather than uncontrolled nucleation. Good seeding transforms crystallisation from an unpredictable event into a reproducible process. But seeding is delicate. Seed too early, and nothing happens. Seed too late, and nucleation has already taken over. Seed with the wrong form or contaminated seeds, and the entire batch may follow the wrong path. Seeding is less a recipe than a negotiation with the molecule, conducted at the right moment and under the right conditions.

Cooling profiles tell their own story. Rapid cooling creates high supersaturation, driving nucleation and producing fine particles that may be difficult to filter, dry, or formulate. Slow cooling encourages growth but may reveal hidden transitions, especially in enantiotropic systems where stability changes with temperature. Antisolvent addition adds further complexity, creating gradients in concentration and localised zones of high supersaturation. Each variable affects the pathway the system follows, and each must be understood rather than assumed.

Then there is morphology, the shape of the crystals themselves. Needle-like crystals may filter poorly or clump together in unpredictable ways. Plates may break easily, generating fines that transform or dissolve at different rates. Compact crystals may be robust but dissolve slowly. Morphology emerges from the interaction between molecular structure, solvent, supersaturation, and process conditions. It is not fixed but shaped by every choice the development team makes.

A well-designed crystallisation process accomplishes several things at once. It defines which solid form is obtained. It delivers particles of the right size and shape. It removes impurities, sometimes acting as one of the most efficient purification steps in the entire synthesis. And it creates a material that can be handled, filtered, dried, milled, blended, compressed, and finally formulated. Crystallisation is not merely a step in the process, it is the bridge between chemistry and pharmaceutics.

The challenge is that crystallisation, more than almost any other operation, exposes the difference between understanding a phenomenon and controlling it. A process that appears reproducible for months may suddenly shift because a parameter previously considered irrelevant becomes influential at scale. Supersaturation may build too quickly. A metastable form may nucleate unexpectedly. A solvate may form under slightly different drying conditions. Equipment geometry may alter mixing patterns. Heat transfer may lag. Each of these changes can redirect the system along a different path.

This is why crystallisation development requires both scientific insight and practical craftsmanship. Models and simulations can help predict trends in supersaturation and growth. Solubility measurements can provide essential boundaries. Mechanistic understanding guides choices about seeding, cooling, and solvent selection. But ultimately, success depends on working closely with the molecule, observing, interpreting, adjusting, and learning from its behaviour.

A robust crystallisation process provides confidence that the material produced tomorrow will behave like the material produced today. It ensures that the chosen solid form persists throughout processing. It protects the molecule from unwanted transitions. And it lays the foundation for every downstream operation. When crystallisation goes well, the entire development journey is smoother. When it goes poorly, formulation becomes harder, manufacturing becomes unpredictable, and surprises appear when the stakes are highest.

In the chapters ahead, we will explore particle size reduction and enabling technologies, tools that complement crystallisation when solubility or dissolution create barriers. But it is worth remembering that the simplest, most elegant solutions often arise when we listen carefully to how the molecule prefers to crystallise and design our processes accordingly.

Crystallisation is where molecules become materials, and where the physical reality of a drug first takes shape.

Chapter 11 — Particle Size Reduction: Engineering Surface Area for Dissolution

Long before the pharmaceutical industry had sophisticated enabling technologies, before amorphous dispersions or lipid systems or cocrystals entered the vocabulary of development teams, there was particle size reduction. Grinding, milling, micronizing... the oldest tools we have for coaxing poorly soluble molecules into something the body can absorb. At first glance, the idea seems almost too simple: if the drug dissolves slowly, make the particles smaller. And indeed, reducing particle size has saved more programmes than almost any other intervention in the history of small-molecule drug development.

But like many simple ideas, particle size reduction reveals its complexity the moment one looks beneath the surface.

Dissolution begins at the boundary where the solid meets the liquid. The greater the area of that boundary, the faster the molecules can escape into solution. Shrinking the particles increases the available surface dramatically. A large crystal may offer only a few square micrometres of interface. Break it into hundreds of smaller fragments, and the collective surface becomes vastly larger. For molecules whose limitation is the *rate* at which they dissolve, this increase in surface area can transform a sluggish curve into one that delivers meaningful exposure.

Yet this elegant principle works only when the true limitation is kinetic. If the molecule is in Class IIa territory, this is, dissolution-limited rather than solubility-limited, then particle size reduction may be enough to shift the development trajectory. But when the molecule sits firmly in Class IIb, where thermodynamics impose a hard ceiling on the maximum achievable concentration, no amount of milling can force more drug into solution. The particles may dissolve faster, but they will still hit the same wall. The mistake arises when teams assume, without evidence, that milling is a universal fix. It is not. It is a tool, powerful in the right context, ineffective in the wrong one.

Even when the physics supports the use of particle size reduction, the process itself introduces new variables. Milling is not a neutral act. It imposes energy on the material, sometimes enough to distort the lattice or induce partial amorphisation. A small amorphous layer on the surface of crystals can temporarily enhance dissolution, but it can also accelerate reactivity or introduce instability. In some cases, milling leads to unwanted transformations, triggering the nucleation of a different polymorph. In others, it creates fines so fragile that they fracture further during processing, complicating downstream operations.

Micronisation magnifies these effects. Jet milling, a common approach, propels particles into one another at high velocity, breaking them through collision rather than shear. It is elegant, solvent-free, and widely used, yet it carries the same risks: heat generation, electrostatic charge, and the potential for structural changes that alter behaviour in unpredictable ways. A compound may appear crystalline after milling, only for subtle DSC or XRPD signals to reveal that the process has softened the lattice or introduced metastable fractions. These details matter, because the behaviour of the milled material is not always the behaviour of the original solid.

Wet milling brings its own nuances. Suspending the drug in a liquid medium reduces heat and protects thermally sensitive materials, but it adds the complexity of drying and the possibility of solvent inclusion. It can produce extremely small particles, even reaching the submicron range, yet it requires careful choice of stabilisers to prevent aggregation. Nanomilling, the most extreme form of size reduction, can produce particles so small that their dissolution characteristics change dramatically. But nanoparticles behave differently from conventional powders. They can agglomerate, change surface charge, or exhibit unexpected stability profiles. Their promise is immense, but they demand understanding and control.

There are also approaches that achieve small particles without mechanical energy. Spray drying can produce amorphous or semi-amorphous fine powders with high surface area. Supercritical fluid technologies can create submicron particles with narrow distributions. These methods, while powerful, may shift the molecule into states where stability, rather than surface area, becomes the dominant concern.

The great paradox of particle size reduction is that its goal is straightforward, more surface area, but its consequences can be noticed across every downstream operation. Fine powders may dissolve beautifully but flow poorly, caking in hoppers or distributing unevenly in blends. They may compress unpredictably into tablets. They may retain moisture differently, heat up during processing, or disperse into the air during manufacture. The very properties that improve dissolution can complicate everything else.

This is why particle size reduction, despite being one of the oldest tools available, must be used with precision. It is most effective when aligned with a deep understanding of the molecule's solubility profile, its polymorphic behaviour, and its mechanical properties. When applied with intention, it can rescue a programme, enabling exposure without resorting to more complex technologies. When applied reflexively, it can obscure the real problem or introduce new ones.

What makes particle size engineering so central to early development is that it sits at the intersection of chemistry, physics, and process. It asks scientists to think about the molecule not as a structure, but as an object with mass, shape, and texture, an object that must dissolve, flow, compress, and survive manufacturing. It challenges teams to see dissolution not as a number on a chart, but as the expression of surface area, solubility, and solid form interacting simultaneously.

There are times when milling is the simplest, most elegant way to unlock a molecule's potential. There are times when it is the wrong tool entirely. The key is recognising which situation you are in, and understanding what surface area can and cannot achieve.

In the next chapter, we turn to another approach, one that often complements particle size engineering when solubility barriers remain too high: the enabling formulations that give molecules a second chance when physics alone is not enough.

Chapter 12 — Choosing the Form the Molecule Can Live With

Polymorphs, salts, cocrystals, crystallization routes, particle engineering... these may appear to be separate decisions, but in practice they converge on a single question: *what solid form allows the molecule to succeed without constant negotiation?*

This part explored the reality that molecules do not always present a single answer. They offer options, some stable, some fragile, some workable only under precise conditions. Solid-form strategy is not about picking the most soluble, the most stable, or the most elegant form. It is about choosing the form that aligns with the real constraints of dose, manufacturing, stability, and formulation.

One consistent theme across projects is that teams often choose too quickly or too narrowly. A form that works for early exposure is sometimes allowed to become the default without asking whether it can withstand scale-up, humidity, drying, processing stresses, or 18 months of storage.

Let the molecule show you its full landscape before committing to a path you cannot easily reverse. Investigate alternatives early. Test assumptions gently. And recognise that the *best* form is rarely the one that looks most attractive in isolation, but the one that behaves faithfully when everything around it becomes more demanding.

PART IV – ENABLING FORMULATIONS FOR PRECLINICAL SUCCESS

Chapter 13 — Early Formulation: Giving Molecules a Fighting Chance

In the early stages of drug development, formulation is often seen as something that comes later, a step that belongs downstream, after the chemistry is settled and the preclinical programme is well underway. But the reality is much less forgiving. The body does not care that a molecule is potent. It does not care how beautifully it binds to its target, or how elegant the synthetic route may be. It cares only about what reaches systemic circulation. And that, in the end, depends on formulation.

Early formulation is where theory meets physiology. It is where the molecule encounters real biological environments, i.e. stomach acid, intestinal fluid, transit times, enzymes, food effects, competing substrates. It is also where the first signs of trouble often appear. A compound that behaves admirably in vitro can falter the moment it is placed in a dosing vehicle. A molecule that looks soluble in a small-volume screen can drop out of solution instantly when diluted in gastrointestinal fluid. A drug with promising rodent data may show erratic behaviour in dogs, revealing weaknesses that were invisible only weeks earlier. These are not failures of biology but mismatches between the molecule's physical properties and the environment it must navigate.

The goal of early formulation is not to build the final dosage form but to give the molecule a fair opportunity to demonstrate its true potential. It is an act of enabling, of clearing the path so that the compound's pharmacology can be evaluated without being obscured by avoidable barriers. A good early formulation does not hide the molecule's weaknesses, but it does not exaggerate them either. It provides clarity.

One of the first decisions teams face is which dosing vehicle to use in preclinical studies. For oral administration, the options range from simple suspensions to more elaborate systems containing surfactants, co-solvents, or lipids. Each vehicle has its own strengths and limitations, and each interacts differently with the molecule. A suspension may be appropriate for a compound with reasonable solubility and dissolution rate, but it will fail if the drug precipitates too quickly or cannot wet properly. A co-solvent system may improve solubility, but it may also create artefacts, such as precipitation upon dilution that leads to variable exposure. Polymeric systems can stabilise supersaturated solutions, but they must be chosen thoughtfully to avoid incompatibilities.

The complexity increases when considering different species. A vehicle that performs beautifully in rodents may be unsuitable for dogs or non-human primates, whose gastrointestinal physiology differs in ways that matter. Differences in pH, bile salt concentration, gastric emptying, and intestinal transit time all influence whether a molecule dissolves, precipitates, or remains available for absorption. A formulation that delivers excellent exposure in one species may underperform in another, not because the molecule has changed, but because the conditions around it have shifted. Early formulation must therefore respect not only the molecule but the biology of the model.

Dose level adds yet another dimension. A compound that performs well at low doses may struggle at higher loads. Solubility limits are exposed, precipitation becomes more likely, the vehicle may no longer carry the drug without reaching saturation. The "100 mg approach" from earlier chapters becomes tangible here: a formulation that is feasible at 10 mg may not be feasible at 500 mg. Early formulation decisions must anticipate not only what is needed today, but what may be needed tomorrow. A good strategy leaves room for dose escalation without collapsing under its own constraints.

Sometimes, the most effective early formulation strategy is not to push the molecule harder, but to step back and reconsider the solid form. A poorly soluble polymorph may need to be replaced with a more soluble one. A salt may offer a better pH-solubility profile. A cocrystal may unlock dissolution behaviour that aligns with the dosing requirements. In other cases, an amorphous approach may be necessary, introducing higher apparent solubility and faster dissolution, at the cost of maintaining physical stability. These decisions are not purely formulation choices, they are solid-state choices expressed through formulation.

What makes early formulation both challenging and indispensable is that it requires scientists to think ahead while improvising in the present. The team may have only grams of material, but must design studies that predict behaviour at much larger scales. There may be limited data, yet decisions must be made quickly because the project cannot wait. There may be uncertainty, yet the formulation must still perform reliably enough to support toxicology, PK, or proof-of-concept studies.

When done well, early formulation acts as both a diagnostic and a bridge. It reveals whether exposure limitations arise from solubility, permeability, metabolism, or dissolution. It shows which directions are promising and which are dead ends. And it buys time for the chemistry to evolve, for the solid form to be refined, or for a more sophisticated formulation to be developed later. It allows teams to progress with confidence rather than hope.

Formulation, in this sense, is not simply a technical discipline. It is a way of listening to the molecule. When a compound refuses to dissolve, it is telling us something. When it precipitates upon dilution, it is telling us something else. When exposure varies between studies, it is sending a message about the limits of the approach. Early formulation is the process of interpreting those messages, understanding what the molecule needs, and giving it the best possible chance to reveal its pharmacological truth.

In the chapters that follow, we will explore how these early decisions connect with more structured strategies, from choosing vehicles across species to managing routes of administration, and later, to the enabling technologies that support molecules when simple approaches are no longer enough. But it all begins here, with the quiet, essential task of giving the molecule a fighting chance.

Chapter 14 — Building Preclinical Dosing Vehicles: Species, Routes, and Real-World Constraints

In the controlled environment of a laboratory, it is easy to imagine that a drug's journey begins only after it reaches human studies. But in truth, that journey begins much earlier, in the quiet and sometimes improvised world of preclinical formulation. This is where the molecule first encounters real biological systems, where dosing becomes physical rather than theoretical, and where constraints reveal themselves in ways that no *in vitro* experiment can fully predict.

Preclinical formulation is, at its core, an exercise in translation. We take the molecule as it exists on the bench, powder, crystals, sometimes amorphous or metastable, and translate it into something an animal model can tolerate, absorb, and metabolise. But unlike clinical formulation, which benefits from established pharmaceutical technologies, preclinical dosing is shaped by constraints imposed by biology rather than by regulations or manufacturing considerations. The volumes are smaller, the flexibility is greater, and improvisation plays a far more central role.

What complicates this work is that each species has its own physiological landscape. A formulation that behaves beautifully in mice may fail entirely in dogs. A vehicle tolerated by rats may cause irritation in the stomach of a non-human primate. The solubility challenges that appear subtle in rodents may become unmanageable at the higher doses required for larger species. And even when the formulation itself remains unchanged, differences in gastric pH, enzyme content, bile salt concentration, and intestinal transit can transform the exposure profile.

Rodents, for example, are forgiving in ways that can be misleading. Their rapid gastric emptying, high metabolic rate, and relatively small doses often allow suspensions or simple vehicles to perform adequately. A compound with only moderate solubility may display acceptable exposure simply because the absolute mass required for dosing is small. It is tempting to feel reassured by these results and to assume that the formulation has passed its first test. But what works at a 1 mg/kg dose in a mouse rarely works at the 20 mg/kg dose required in dogs or monkeys. The formulation that once felt robust becomes fragile the moment dose escalates.

Larger species confront us with realities that small ones conceal. Dogs, for example, often expose the limits of solubility with brutal honesty. Their stomach pH can fluctuate dramatically, sometimes from acidic to nearly neutral depending on feeding state, causing compounds to dissolve in one moment and precipitate the next. Their dosing volumes are larger, but not infinite. And their gastrointestinal tract, with its longer transit times, often magnifies dissolution limitations that rodents seem to forgive.

Non-human primates add their own complexities. Their gastric and intestinal environments resemble those of humans more closely, but their tolerability to excipients is often lower. Co-solvents that pose no issue in rodents may cause gastrointestinal distress in primates. Surfactants that are well tolerated by dogs may be unsuitable. Even simple oils can behave differently, forming emulsions or precipitates that derail absorption. Working with primates requires a delicate balance between enabling solubility and protecting tolerability, and this balance is rarely straightforward.

Across all species, volume is one of the most defining constraints. Every route of administration has its limits. Oral dosing volumes cannot be infinitely increased; stomachs can hold only so much. Intravenous dosing has even stricter limits—especially in small animals—forcing formulators to work within narrow solubility windows. Subcutaneous and intraperitoneal routes introduce their own ceilings, dictated by local irritation and tissue capacity. Even when a vehicle can solubilise the dose in theory, the volume required may exceed what the animal can physically tolerate. These constraints reshape the problem, turning a solubility challenge into a feasibility challenge.

The tools available to preclinical formulators are diverse: co-solvents, surfactants, oils, polymers, pH adjustments, complexation systems; but none of them behave identically across species. Ethanol may aid solubility but irritate the stomach. PEG may hold the drug in solution but alter osmolarity. Surfactants like Tween or Cremophor can enhance dissolution but risk altering permeability or causing unexpected interactions with the gastrointestinal environment. Cyclodextrins can provide elegant solutions but only within limits imposed by their own tolerability profiles. Every choice carries its own risks and its own story.

And so, developing a preclinical vehicle becomes an act of constraint navigation. You begin with what the molecule needs to dissolve and what the species can tolerate, then search for the narrow space where those two requirements overlap. You consider the dose, the route, the behaviour of the solid form, the risk of precipitation upon dilution, and the physiology of the animal. You learn to recognise early signs of incompatibility, cloudiness that hints at impending precipitation, viscosity that signals an unmanageable suspension, or a surfactant level that pushes the limits of tolerance.

Perhaps the most important skill in preclinical formulation is the ability to look beyond the numbers. A solubility of one milligram per millilitre may seem acceptable until you calculate the required dose for a dog. A clear solution may inspire confidence until you dilute it in simulated intestinal fluid and watch it crash out in seconds. A beautiful suspension may appear stable until you realise that half the dose has adhered to the dosing syringe. The science is quantitative, but the judgement comes with experience.

What makes preclinical formulation so valuable is that it forces the molecule to reveal its behaviour early. It exposes weaknesses in solubility, dissolution, and stability. It highlights the role of particle size, the tendency to precipitate, the sensitivity to pH, and the risks of conversion. It prepares the team for the realities of clinical formulation, sometimes confirming that the chosen solid form is adequate, and sometimes signalling that a deeper transformation, salt, cocrystal, amorphous, or enabling technology may be necessary.

Above all, preclinical formulation is a reminder that molecules must exist in a world that does not bend to their preferences. Biology is indifferent. Physiology imposes limits. And the molecule, with all its strengths and vulnerabilities, must work within those constraints. Understanding this early makes everything that follows more predictable.

In the next chapters, we will explore enabling technologies, not as silver bullets, but as thoughtful interventions for molecules that need more help than simple dosing vehicles can provide. But the foundation of all of these strategies lies in what we learn here, in the pragmatic, constraint-driven world of preclinical formulation.

Chapter 15 — From Possibility to Exposure: The Real Purpose of Early Formulation

Early formulation often feels provisional. Temporary vehicles, improvised strategies, gram-scale constraints... but beneath the improvisation lies a simple, decisive purpose: to reveal the molecule's true biological potential without being misled by avoidable physical limitations.

This part highlighted a reality familiar to anyone who has sat through preclinical PK reviews: exposure is the first real performance test of the molecule, and formulation is the lens through which that test is interpreted. A vehicle that masks precipitation, or a suspension that wets poorly, or a solubilising system that collapses upon dilution... all of these shape the conclusions a team draws about potency, viability, and dose.

Design early formulations not as placeholders, but as diagnostic tools. The right vehicle tells you whether solubility is the bottleneck, whether dissolution is limiting, whether supersaturation holds, whether the solid form is cooperating. The wrong vehicle tells you very little.

Good early formulation does not rescue a molecule, it clarifies it. And clarity is what allows a project to move forward with confidence rather than improvisation.

PART V – STRATEGY, IP, AND THE BUSINESS OF SOLID STATE

Chapter 16 — The Business Value of Solid-State Decisions

If there is one truth that becomes clear after years of working with development teams, it is that scientific decisions rarely stay confined to the laboratory. The choice of a solid form, the way a crystallisation process behaves, or the performance of an early formulation can ripple outward, influencing timelines, budgets, valuations, negotiations, and even the commercial trajectory of a drug. Science and business are not separate worlds; they are deeply entangled. And nowhere is this more apparent than in the solid state.

Early in development, teams often make decisions with a narrow focus: What do we need today? Can we get enough exposure for the next PK study? Can we produce material for tox? Can we move forward without delay?

These questions are legitimate, every programme faces time pressure, but they can obscure the long-term implications of early choices. A solid form selected hastily for convenience may prove difficult to reproduce at scale or may transform under processing conditions. A formulation designed as a temporary fix may become entrenched, shaping expectations and complicating later transitions. A poorly understood solubility profile may lead to unstable dose projections that disrupt forecasting and investor communication. Each early assumption, if made without a full understanding of the solid state, becomes a future liability.

For small biotech companies, the business consequences are particularly acute. Their value is often tied not to revenue, but to potential, an intricate calculation of scientific promise and development risk. Investors, licensing partners, and acquirers all assess that risk in different ways, but one factor they evaluate with increasing scrutiny is developability. A molecule with unknown polymorphic behaviour, poorly defined solubility, or unstable formulations carries hidden risks that can reduce upfront payments, lower deal valuations, or complicate negotiations. Conversely, a programme that demonstrates a clear understanding of the solid form, a robust crystallisation route, and a predictable exposure profile carries significantly more weight.

The difference can be dramatic. I have seen programmes stagnate because exposure could not be achieved consistently. I have seen partnerships delayed because the manufacturing team could not produce the same solid form twice. I have seen promising molecules rejected because of dose projections that were incompatible with oral delivery. These are not strictly scientific failures; they are failures of risk communication. When a partner senses uncertainty in the fundamentals of the molecule's physical behaviour, confidence erodes. And when confidence erodes, so does value.

Solid-state clarity has the opposite effect. When a biotech can show that the polymorphic landscape has been mapped, that the chosen form is stable, that the crystallisation process is reproducible, and that formulation strategies align with the molecule's properties, partners respond differently. Risk is no longer a black box, it becomes something quantifiable, manageable, and therefore negotiable. The science does more than solve technical problems, it strengthens the business case.

The same principles apply, in a different way, to generic manufacturers. In that world, the solid state is not merely a source of risk, it is a source of opportunity. The choice of a particular polymorph, hydrate, salt, or cocrystal can determine whether a generic product infringes a patent, whether it circumvents one, or whether it occupies a legally defensible middle ground. Entire generic strategies have been built on the ability to produce a form the innovator did not describe, or on the recognition that the marketed form is not the only possible one.

But these opportunities require deep understanding. A generic version that relies on a metastable polymorph may face stability issues that jeopardise bioequivalence. A salt form may disproportionate during storage. A hydrate chosen to avoid a patent may convert to another form under manufacturing conditions. The competitive advantage offered by the solid state becomes meaningful only when the chosen form behaves predictably across the entire lifecycle of the product.

Whether in innovative development or generics, the business value of solid-state decisions comes from their ability to reduce uncertainty. Uncertainty in exposure, uncertainty in manufacturing, uncertainty in stability, uncertainty in cost, uncertainty in IP. The fewer surprises a molecule delivers, the more attractive it becomes, not only scientifically but strategically.

This is why companies that invest early in understanding the solid state often find themselves moving faster later. They avoid the delays associated with unexpected transitions, costly reformulations, or last-minute rescue strategies. They present clearer development plans to partners. They negotiate from a position of strength because they can articulate the physical behaviour of their molecule with confidence. And perhaps most importantly, they preserve optionality, the freedom to choose the most appropriate formulation, the best manufacturing route, or the strongest IP position.

There is a quiet irony here. Many people see solid-state science as a technical speciality, far removed from commercial strategy. But in practice, it is one of the most strategic disciplines in early development. It determines how smoothly a molecule progresses, how predictable its behaviour is, how costly its development will become, and how attractively it can be positioned in the eyes of those who might license, invest in, or eventually compete with it.

In the end, the business value of solid-state decisions lies not in avoiding problems but in understanding them early enough, so they never become crises. A molecule that behaves well in the solid state is easier to formulate, easier to manufacture, easier to scale, and easier to explain. And in drug development, ease is a form of value.

In the next chapters, we will turn to intellectual property and case studies, examples where the solid state has shaped not just science, but entire commercial outcomes. These stories illustrate, often vividly, how deeply the architecture of a molecule influences the business built around it.

Chapter 17 — Intellectual Property in the Solid State: Lessons from Paroxetine, Gatifloxacin, and the Courts

For many people outside the pharmaceutical world, intellectual property conjures images of chemical structures, synthetic routes, and novel mechanisms of action. But inside the industry, intellectual property extends far beyond the structure of a molecule. It inhabits the solid form: the way the molecules pack, the presence of water in the lattice, the counterion chosen for a salt, the architecture of a cocrystal. In other words, the very materiality of the drug becomes patentable.

This makes the solid state an arena where science and law intersect in ways that are sometimes elegant, sometimes combative, and occasionally controversial. Over the last decades, courts around the world have shaped the rules of what can and cannot be protected. At the centre of many of these decisions lie stories of discovery, dispute, and unexpected scientific behaviour, stories that remind us how powerful, and how precarious, solid form patents can be.

Few examples illustrate this better than paroxetine.

Paroxetine was originally marketed as a hemihydrate. It worked well, but as with many successful drugs approaching the end of their patent life, the innovator sought ways to extend exclusivity. The solution seemed straightforward: develop and patent a new crystalline form. SmithKline Beecham discovered a new non-solvated form, characterised it, demonstrated differences, and secured patents. On paper, it appeared to be a defensible extension.

The legal battle that followed changed how courts view solid form patents. When challenged, it became evident that the non-solvated form reverted to the hemihydrate in vivo, quite literally inside the patient. The solid form that entered the mouth was not the solid form that entered systemic circulation. In a case that has since become part of pharmaceutical folklore, this conversion was demonstrated using what became known as the "*string test*": a tablet tied to a thread, swallowed, retrieved, and analysed. Simple, almost absurd in its imagery, yet scientifically unambiguous.

The courts concluded that if the new form did not confer a meaningful therapeutic difference, and if its discovery was within the routine experimental capabilities of the field, it did not meet the bar for inventiveness. The patent fell. And with it, the industry received a clear message: to protect a new crystalline form, it is no longer enough to show that it exists. You must show that it matters.

Gatifloxacin tells a different kind of story, one less about litigation and more about the sheer unpredictability of solid-state landscapes. What began as a straightforward hydrate became, through systematic crystallisation studies, a family of more than a dozen forms, each with its own quirks. Some offered stability, others solubility, others manufacturability. Some appeared only under narrow conditions, others emerged unexpectedly when drying protocols changed. The development programme was delayed not because of a lack of innovation, but because the molecule refused to fit neatly into a single structural identity.

From an IP perspective, Gatifloxacin illustrates both the opportunity and the risk. A rich polymorphic landscape can offer multiple patentable forms. But it can also make it harder to claim inventiveness if the molecule readily forms many structures with minimal prompting. The more easily a compound crystallises into alternative arrangements, the more a court may view those arrangements as predictable rather than inventive. And unpredictability, paradoxically, can sometimes weaken a claim rather than strengthen it.

Then there is ritonavir, perhaps the most infamous case of all, not because of litigation, but because the molecule itself rewrote the rules. The sudden appearance of a new, more stable polymorph after commercial launch was a scientific shock that reverberated through the entire industry. The issues were not legal but existential. A previously unknown form had become the only form the molecule seemed willing to adopt. Manufacturing halted. The drug was reformulated. Patients were affected. Millions were lost.

From an IP standpoint, ritonavir changed the way regulators and companies viewed solid forms. It became clear that if even a fully commercialised drug could produce a new polymorph years after launch, then polymorphism must be treated as a fundamental risk, not an afterthought. Companies began investing more aggressively in polymorph screening, not only to find new forms that could be patented, but to avoid the disaster of discovering them too late. The boundary between IP strategy and scientific risk management blurred, they became two sides of the same coin.

What unites these cases is not the specific molecules or the outcomes, but the lesson that solid-state IP lives in a delicate space. Too predictable, and the courts will dismiss it as obvious. Too unpredictable, and the molecule may undermine its own protection. Too subtle, and the claimed advantages may be dismissed as trivial. Too broad, and the patent may fail for lack of enablement. Solid-state patents must thread the needle between novelty, utility, and non-obviousness, and doing so requires not only scientific rigour but a deep understanding of how courts interpret scientific behaviour.

This is why successful solid-state IP strategies rarely focus solely on one form. They involve understanding the entire landscape—polymorphs, hydrates, solvates, salts, cocrystals—and identifying which forms offer meaningful, defensible advantages. They require evidence that those advantages matter in practice: improved stability, better bioavailability, superior manufacturability, or clear therapeutic relevance. And they benefit from alignment between scientific development and IP timing, ensuring that discoveries are supported by data robust enough to stand up under scrutiny.

For innovator companies, strong solid-state IP protects investment, strengthens partnering opportunities, and extends commercial life. For generic manufacturers, understanding the landscape provides pathways for differentiation and competition. For both, the solid state becomes not just a scientific domain but a strategic one.

In the end, the stories of paroxetine, gatifloxacin, and ritonavir are not only case studies in IP, they are reminders of how deeply the physical behaviour of a molecule can shape its legal, commercial, and therapeutic destiny. A drug's success is written not only in its chemistry, but in the way its molecules choose to assemble themselves. And the law, like the science, must adapt to that complexity.

In the next chapter, we will step further into the future, exploring how computation, artificial intelligence, and predictive modelling are beginning to reshape how we discover, screen, and select solid forms, and what that means for the next generation of medicines.

Chapter 18 — The Future: AI, Predictions, and the Next Era of Solid-State Science

For most of the history of solid-state chemistry, progress has been driven by careful experimentation, informed intuition, and an intimate familiarity with how molecules behave in the real world. Crystallisation was learned through practice. Solid form selection relied on experience as much as on theory. And while computational chemistry contributed insight, the solid state remained a domain where experiments carried the final word.

That is still true today, but the balance is shifting.

Over the past decade, computational tools have begun to step into a more prominent role. What started as an aspiration, predicting physical properties before synthesising a molecule, is slowly becoming feasible. Artificial intelligence, machine learning, and crystal structure prediction no longer feel like distant promises, they are becoming part of the everyday vocabulary of development teams, at least in companies bold enough to integrate them early.

The excitement around these tools is understandable. Polymorphs, cocrystals, and other solid forms are notoriously difficult to predict using intuition alone. The landscape is broad, the number of possible arrangements enormous, and the variables too interdependent for simple models. When experimental screens reveal unexpected forms, it is often not because the scientists failed, but because the molecule explored a region of the landscape no one anticipated.

Computational approaches offer something new: the possibility of mapping that landscape before setting foot in the lab.

Crystal structure prediction (CSP) has grown from a specialised academic pursuit into a tool with genuine industrial relevance. While CSP cannot yet guarantee that every possible polymorph will be predicted, it can often identify the most likely low-energy packing arrangements, hinting at which forms are stable, which are metastable, and which combinations are energetically plausible. For salts and cocrystals, CSP can estimate lattice energies, helping teams prioritise combinations that are not only chemically reasonable but physically robust.

Machine learning adds another layer. Instead of solving equations that describe intermolecular interactions, ML learns from vast datasets of structures, energies, solubilities, melting points, crystallisation outcomes. It recognises patterns that no human could memorise and uses them to predict properties of new molecules. These tools can suggest promising coformers for a cocrystal, estimate solubility trends, identify likely hydrogen-bonding motifs, or flag structural features associated with metastability. As the datasets grow, the accuracy improves.

Yet, despite these advances, computational tools are not replacements for experimental work. They are guides, sometimes brilliant, sometimes imperfect, occasionally misleading. They narrow the search space, but they cannot eliminate uncertainty. They can rank possibilities, but they cannot guarantee which form will appear under real-world conditions, with real solvents, under real process stresses. They are models of the landscape, not the landscape itself.

What makes the future so interesting is the growing synergy between prediction and experiment. In the past, experimental screens were broad, exploratory, and sometimes unfocused. Today, computational tools can direct that effort, pointing toward the most promising regions of crystallisation space or the coformers most likely to succeed. Instead of hundreds of experiments, teams may need only dozens, targeted, informed, and far more efficient.

Some companies are already embracing this approach. They combine CSP with ML-driven coformer selection, integrate predictive models into recrystallisation workflows, and use automated platforms to test conditions guided by computational insight. The result is not a shortcut but a partnership: machine-guided hypotheses refined by experimental evidence, and experimental data feeding back to improve the models.

The biggest gains may come not from predicting the structure itself, but from predicting risk. If a computational screen suggests that a more stable polymorph is energetically accessible, teams can plan for that possibility long before it emerges. If an amorphous material shows a high likelihood of recrystallisation, formulators can choose polymers more intelligently. If a cocrystal appears promising, CSP can help determine whether its lattice is strong enough to survive scale-up. These insights make development more strategic, not more rushed.

Still, there are limitations we must acknowledge. CSP struggles with flexible molecules and large conformational spaces. ML is only as good as the data it learns from, and much of the pharmaceutical world's data remains proprietary or unpublished. Neither approach can fully account for impurities, solvents, mechanical stress, or the quirks of crystallisation kinetics. The molecule's behaviour in a reactor is still something only experiments can reveal.

But if the solid state has taught us anything, it is that understanding emerges gradually. A tool does not need to be perfect to be transformative, it needs only to reveal part of the picture more clearly than before. AI and computational methods already do this. They bring structure to uncertainty. They help us ask better questions before approaching the bench. They turn random exploration into guided discovery.

The next era of solid-state science will not be fully computational and will not be fully experimental. It will be both. The best scientists will be the ones who can interpret predictions without being blinded by them, who can design experiments that test not only what is likely but what is possible, and who can use insight from both worlds to create a deeper, more unified understanding of the molecule.

The future will not eliminate surprises. Polymorphs will still appear unexpectedly. Cocrystals will still challenge our intuition. Amorphous systems will still defy simple rules. But the surprises will become fewer, and they will occur earlier in development, when they can still be managed without crisis.

And perhaps that is the most important point: computation does not remove uncertainty, it simply moves it upstream, where it can be handled with far greater confidence. It gives us foresight, not knowledge. It strengthens our questions, not our assumptions.

In the final chapters, we will turn to real-world stories, i.e. ritonavir, gatifloxacin, paroxetine, and see how different the outcomes might have been should today's tools, thinking, and integration of computational insight had existed then. These stories are reminders not only of how far we have come, but of how much more we now understand about the quiet, powerful world of the solid state.

Chapter 19 — When Physical Behaviour Becomes Strategy

Science and business are often treated as separate tracks in drug development, but anyone who has watched a programme stumble because of a solid-form issue knows they are inseparable. In this part, the thread connecting decisions, value, timelines, IP, and risk became visible.

Solid-state behaviour is not merely a scientific characteristic; it is a commercial factor. A stable form accelerates manufacturing and builds confidence. An unstable one erodes timelines and negotiating power. A defensible cocrystal may extend exclusivity, while a predictable hydrate may not. A formulation that performs reliably becomes an asset, but one that varies undermines valuation.

Treat solid-state understanding as a value-generating activity, not a technical checkbox. Every insight reduces uncertainty. Every well-chosen form protects variability. Every avoided crisis preserves time, cost, and credibility.

In a world where programmes move quickly and right decisions matter, the solid state quietly shapes outcomes long before anyone realises it. Seeing it early is not caution, it is advantage.

PART VI – LESSONS FROM REAL-WORLD CASES

Chapter 20 — Ritonavir: When a Molecule Rewrites the Rules

Every discipline has its defining moment, an event so unexpected, so disruptive, that it changes the way people think forever. In solid-state chemistry, that moment arrived in 1998, when ritonavir, a life-saving HIV drug, revealed a truth that had always been present but never fully confronted: molecules can change their mind.

When ritonavir first entered the market, it arrived as a triumph of medicinal chemistry and formulation ingenuity. Protease inhibitors had transformed the outlook for patients with HIV, and ritonavir was a cornerstone of that progress. It had been studied, manufactured, validated, approved, and launched with confidence. The solid form was known. The formulation was known. The manufacturing process was controlled. Everything appeared stable.

Then, without warning, the drug stopped behaving.

Batches that had been manufactured reliably for years suddenly failed dissolution testing. Capsules that once held a homogeneous semisolid material now contained visible crystals. The active ingredient, once fully soluble in the formulation, began precipitating out. It wasn't a minor deviation, it was a collapse. Ritonavir, the same molecule that had sailed through years of development, now existed in a form that no one had seen before.

What emerged was a new, more stable polymorph. One that had never been detected during development, never appeared in thousands of batches, and, most unsettlingly, seemed to propagate itself. Once this new form surfaced, it became increasingly difficult to prevent it from appearing again. Seeds invisible to the eye were enough to redirect crystallisation. Processes that had once been robust were now unreliable. What had been a well-understood drug substance had become unpredictable.

For the scientists tasked with understanding what happened, the experience must have been both bewildering and humbling. The investigation revealed that ritonavir was polymorphically richer than anyone realised. The original form, Form I, was not alone. A second form, Form II, existed, more stable, more tightly packed, and far less soluble. Once it appeared, Form II became the energetically preferred structure, meaning the molecule would gravitate toward it whenever conditions allowed. In the manufacturing environment, despite tight controls, conditions did allow it.

With the emergence of Form II, the original formulation could no longer keep the drug in solution. The semisolid capsule was pushed beyond its limits, precipitation became inevitable. Bioavailability plummeted. And because the drug was already on the market, there was no time for gradual adjustment or quiet reformulation. Production had to stop. Supply was disrupted. Patients were affected. The company faced a crisis that no one had imagined possible.

The ritonavir incident forced the entire industry to confront a difficult truth: even late in development, even after approval, even after launch, new solid forms can still appear. Stability is not a promise, it is a condition, one that must be actively understood and controlled.

In retrospect, ritonavir taught us several lessons that are now foundational to solid-state development. The first is that polymorph screening must be thorough, rigorous, and repeated under diverse conditions. Not because we expect to find every possible form, no one can guarantee that, but because understanding the landscape reduces the risk of being blindsided.

The second lesson is that process itself can reveal new physics. The laboratory may show one behaviour, but manufacturing shows another. Supersaturation, cooling profiles, solvent composition, equipment surfaces... all of these can open polymorphic pathways that never appear at small scale. This is not a failure of chemistry, it is the nature of crystallisation.

The third lesson is that once a more stable form appears, it often becomes extraordinarily difficult to eliminate. In ritonavir's case, Form II did not merely show up, it became dominant. This phenomenon is called *"cross-contamination"* in an everyday sense, but in scientific terms it is the introduction of a lower-energy template into the environment. Once present, those seeds make it easy for the system to fall into the new minimum. It is a reminder that solids, like ideas, spread.

Perhaps the most important lesson of all is that molecules behave according to their own energetic logic, not our assumptions. We can measure, predict, model, and control, but we cannot impose a structure that nature finds less favourable. The best we can do is understand the landscape deeply enough that we avoid stepping into the same traps.

It is no exaggeration to say that ritonavir reshaped regulatory expectations worldwide. After 1998, agencies began demanding far more comprehensive solid-state characterisation, more rigorous control strategies, and deeper justification for the chosen solid form. Industry responded by strengthening screening protocols, integrating crystallisation development earlier, and recognising that polymorphism is not an inconvenience but a fundamental feature of molecular behaviour.

But beyond the scientific and regulatory consequences, ritonavir left a philosophical legacy. It reminded us that molecules are dynamic, responsive, and capable of surprising even the most experienced teams. It humbled us, and humility is often the beginning of better science.

Had today's computational tools existed at the time, this is crystal structure prediction, machine-learning-assisted screening, automated crystallisation platforms, etc., the story might have unfolded differently. Perhaps Form II would have been identified earlier. Perhaps its stability would have been predicted. Perhaps the risk could have been managed before it became a crisis. We cannot know for certain. But we can say that ritonavir accelerated the evolution of solid-state science, pushing it toward deeper inquiry and more proactive thinking.

Every generation of scientists inherits a cautionary tale. In our field, ritonavir is that tale. Not because it represents failure, but because it illuminates the delicate balance between what we know and what molecules are still capable of teaching us.

In the next chapter, we turn to another story, one less dramatic but equally revealing, where a molecule's unexpected behaviour reshaped an entire development strategy: the case of gatifloxacin.

Chapter 21 — Gatifloxacin: Stability, Solubility, and the Surprises in Between

If ritonavir is remembered for its drama, gatifloxacin is remembered for its depth. Where ritonavir shocked the industry with a sudden, late-appearing polymorph, gatifloxacin challenged scientists in a quieter, more persistent way. Its story is not one of crisis, but of discovery, an unfolding complexity that forced its development team to confront just how rich and unpredictable the solid-state landscape of a single molecule can be.

Gatifloxacin began as many compounds do, crystallising first as a hydrate. In this case, a hemihydrate. It was a natural first form: stable enough to handle, easy to reproduce, and well-suited for early studies. That stability, however, came with a cost: the hemihydrate was hygroscopic. It absorbed moisture readily, which made its physical properties sensitive to humidity and complicated both processing and formulation. Tablets made from the hemihydrate tended to disintegrate unpredictably, and dissolution performance varied more than anyone liked.

The discovery of a sesquihydrate seemed promising. This new form possessed more controlled hydration, better stability under ambient conditions, and improved manufacturability. When it was licensed for further development, the sesquihydrate became the lead form. It behaved well, or at least better, and there was optimism that it could bring more predictability to the manufacturing process.

Then came the pentahydrate.

The pentahydrate was, in many ways, the opposite of what the development team expected. It formed readily under certain conditions, and it possessed impressive physical robustness. Its stability was excellent; its crystal packing was orderly and strong. It seemed, structurally, like an ideal form. But stability, as gatifloxacin taught us, is only one dimension. The pentahydrate was significantly less soluble than the earlier forms, and that change in solubility translated directly into lower bioavailability.

The surprise was not simply that a less soluble form existed, but that it appeared in the first place. To understand why, the team began exploring crystallisation conditions more broadly: different solvents, drying protocols, antisolvents, cooling rates, humidity exposures, and processing stresses. What they discovered was staggering. The molecule did not offer three or four possible solid forms. It offered dozens. With nothing more than ethanol, water, and variations in drying, they uncovered more than a dozen additional forms, hydrates, solvates, and polymorphs, each with its own signature.

This was not a molecule defining itself, it was a molecule exploring its possibilities.

The discovery was both scientifically fascinating and practically daunting. Every new form required characterisation. Every form needed to be evaluated for stability, solubility, manufacturability, and compatibility with the intended dosage form. Every new structure introduced new questions: *Would it appear spontaneously? Would it transform into something else during processing? Would it persist during storage? Would it affect bioavailability?*

Development slowed not because anyone made a mistake, but because the molecule demanded a deeper understanding.

Gatifloxacin reminded developers of a truth that is often forgotten: the goal is not merely to find a solid form, but to choose one that remains faithful during scale-up. A form that is beautiful on the bench but unstable under industrial conditions is not an asset, but a liability. A form that appears rare at first glance may become dominant under certain process conditions. A hydrate that behaves predictably at controlled humidity may behave erratically in real-world environments.

In the end, despite the intricacies of the polymorphic landscape, the marketed versions of gatifloxacin were built around forms that balanced solubility, stability, and manufacturability. But the story did not end there. When patents expired, generic manufacturers revisited the landscape with fresh eyes. Interestingly, one of the original forms, the hemihydrate, became the form of choice for at least one generic version. It was easier to produce, easier to handle, and aligned well with regulatory expectations. The challenges that once weighed against it became manageable with minor technological updates. What was once rejected became, in a different context, practical.

There is something elegant about this reversal. It captures the essence of solid-state development: that the *"best"* form is never an abstract ideal, but the one that fits the realities of the moment, the biology, the process, the formulation, the equipment, the regulatory landscape, and the commercial strategy.

Gatifloxacin's polymorphic richness teaches us that molecules are capable of far more variation than we expect. It shows how development teams can lose time not because they miscalculated, but because the molecule offers too many possibilities. And it serves as a reminder that thorough solid-state exploration is not a luxury, but a necessity, not because regulators demand it, but because molecules can surprise us in ways that matter deeply to patients, timelines, and commercial viability.

In the next chapter, we will examine a very different kind of solid-state story, one where the molecule itself became the centre of a legal battle: the case of paroxetine, and what it revealed about inventiveness, obviousness, and the limits of solid-state IP.

Chapter 22 — Paroxetine: When the Solid State Meets the Law

If the story of ritonavir revealed how a molecule can disrupt manufacturing, and the story of gatifloxacin showed how it can complicate development, the case of paroxetine demonstrates something different: how a molecule can shape legal doctrine. It is one of the clearest examples of how science, law, and commercial strategy intersect, sometimes in ways that no one anticipates.

Paroxetine was originally developed, manufactured, and marketed as paroxetine hydrochloride hemihydrate. This was the solid form behind Seroxat/Paxil, and it behaved well: stable enough to handle, predictable, and suitable for large-scale manufacture. When the original compound patent approached expiry, the innovator, SmithKline Beecham, sought to strengthen their exclusivity by identifying and patenting a new solid form of the hydrochloride salt. They found a paroxetine hydrochloride anhydrate. It was distinct, characterisable, and reproducible under the right conditions. In the context of the time, patenting a new solid form of an existing salt was a logical and widely accepted extension strategy.

The legal battle that followed would change expectations for years.

When Apotex challenged the patent, their argument centred on two main points. First, they claimed the discovery of the anhydrate form was obvious, that a skilled solid-state scientist would expect multiple forms of a hydrochloride salt to exist and could discover them using routine experimentation. Second, and far more damaging to the patent's validity, they demonstrated that the anhydrate form did not remain anhydrate *in vivo*. Once the anhydrate tablet was swallowed, it converted into the hemihydrate inside the patient's gastrointestinal tract.

This conversion was not abstract speculation, it was shown experimentally. In what has become an almost legendary example of practical ingenuity, a tablet of paroxetine hydrochloride anhydrate was tied to a thread, swallowed, retrieved, and analysed. The solid form that was swallowed was the anhydrate. The solid form retrieved was the original hemihydrate. Moisture and gastric conditions had reversed the solid-state transformation, erasing the distinction that formed the basis of the new patent.

The court's response was sharp and clear. If the new form did not persist through dosing, if it offered no practical therapeutic advantage, and if its discovery required nothing beyond the standard tools of a skilled scientist, then it did not meet the threshold for inventiveness. The patent was invalidated. The message to industry was unmistakable: a new solid form of a salt is not inherently an invention, it must matter in a meaningful way.

This decision reshaped how companies approach solid-form patents. For years, identifying a new polymorph or hydrate of a hydrochloride salt was often enough to justify additional protection. Paroxetine changed the calculus. Courts began demanding evidence of unexpected properties, this is improvements in stability, manufacturability, solubility, or therapeutic relevance. They looked not just at whether the form existed, but at whether it made a genuine, non-obvious contribution.

The paroxetine case also highlighted a critical scientific point that resonates today: not all solid forms survive the journey from the bottle to the bloodstream. Some retain their structure throughout dissolution. Others convert instantly. If a solid form is chemically or structurally destined to transform under physiological conditions, then its relevance, both scientific and legal, must be questioned.

The consequences extended beyond innovators. Generic manufacturers took notice. The decision encouraged them to investigate the polymorphic landscape of established hydrochloride salts more aggressively, to challenge patents more strategically, and to use solid-state chemistry as an avenue for competition rather than compliance. A solid form became not only a scientific entity, but a strategic one.

What makes the paroxetine case so enduring is that it captures the tension between scientific possibility and legal interpretation. Solid-state chemistry tells us that molecules can adopt many forms. Patent law tells us that not all differences count. Paroxetine sits precisely at that intersection, reminding us that the importance of a salt form is not defined solely by its structure, but by its behaviour in the plant, in the formulation, and, ultimately, in the patient.

Together with ritonavir and gatifloxacin, paroxetine completes a trio of lessons that anchor the science and strategy of the solid state. These cases show that molecules do not simply challenge our manufacturing lines or our analytical tools. They challenge our assumptions, our development strategies, and, occasionally, the boundaries of patent law itself.

Chapter 23 — What the Molecules Tried to Teach Us

The real-world cases in this section are not anecdotes from another era. They are reminders of how molecules behave when we stop paying attention, or when we underestimate the landscapes they inhabit.

Ritonavir taught us that new forms can emerge even when we believe the landscape is known. Gatifloxacin showed us that complexity itself can become the challenge. Paroxetine revealed that solid-state behaviour has legal and commercial consequences as profound as its scientific ones.

Together, these stories point to a single truth: molecules are consistent, even when surprising. They behave according to physics, not expectations. And the issues we consider *unexpected* are often the ones that would have been visible had we looked earlier, or from a slightly different angle.

Treat surprises as signals, not anomalies. They are the molecule's way of telling us where to look next, what assumption to revisit, or which path requires more care.

When we listen, the surprises become far less dramatic. And when we anticipate them, development becomes far more predictable.

PART VII – FUTURE DIRECTIONS AND FINAL THOUGHTS

Chapter 24 – When to Seek Expert Help: Timing, Blind Spots, and Avoiding the Avoidable

One of the most quietly consequential decisions in a drug development programme is not about chemistry or formulation or even the choice of solid form. It is about timing. Specifically: *when should a team seek expert help in solid-state science?*

Ask this question too early, and it may feel unnecessary. Ask it too late, and the consequences can be measured in months, millions, and missed opportunities.

Every development team wants to believe they can manage the molecule on their own. This is natural. Medicinal chemists understand solubility. Formulators understand dissolution. Process chemists understand crystallisation. Analytical scientists understand structure and stability. Each discipline brings deep knowledge, and early on, that knowledge is often enough.

But the moment a molecule transitions from the simplicity of early research into the complexity of real-world development, the challenges begin to multiply. The solid state touches everything:

solubility, bioavailability, manufacturability, stability, process control, regulatory documentation, and intellectual property. No single person, no matter how experienced, holds all the necessary expertise across these domains. Blind spots appear not because teams lack skill, but because the molecule moves into a space where multiple disciplines overlap.

The most common sign that expert support is needed is surprisingly subtle: inconsistency. A solubility measurement that varies more than expected. A crystallisation that works one day and misbehaves the next. A suspension that delivers exposure in one study but not another. A solid form that appears stable on paper yet shows hints of transformation under stress. These variations are often dismissed as noise, but they are usually the molecule signalling that something deeper deserves attention.

Another sign is urgency. A preclinical study approaches, exposure is insufficient, and the team must find a way to deliver a higher dose. A manufacturing process must scale, but the solid form has not been fully understood. A regulatory submission looms, and the supporting data around the solid state feel thin. Urgency compresses time, and compressed time amplifies risk. This is when problems hidden earlier in the project become impossible to ignore.

But the most dangerous situation, and the one solid-state experts encounter most frequently, is when a team has already fallen into a trap without realising it. They have chosen a solid form because it was convenient, not because it was appropriate. They have scaled a crystallisation before understanding its thermodynamics. They have developed a formulation that holds only under narrow conditions. They have assumed a form is stable because it has *behaved* stable, without testing the environments that matter. By the time the warning signs appear, the strategy is already built around these assumptions, making a course correction painful.

This is not a failure of competence. It is a feature of drug development. The earliest decisions are often made with the least data, and the risks only reveal themselves gradually. What solid-state specialists bring to a programme is not merely technical expertise, but perspective, the ability to recognise at a glance what patterns tend to precede trouble.

Sometimes the issue is a hidden polymorph waiting to appear at scale. Sometimes it is a hydrate concealed as an anhydrate. Sometimes it is a solubility limitation that no amount of milling can solve. Sometimes it is a formulation strategy that will collapse when the dose increases. And sometimes it is simply a lack of alignment between solid form, formulation, and process, three pillars that must support each other if a drug is to progress smoothly.

The teams who benefit the most from early expert involvement are not those with the worst problems, but those with the least ambiguity. When a molecule's behaviour is clearly understood early on — its solubility limits, polymorphic tendencies, crystallisation preferences, humidity sensitivity — development becomes more predictable. Surprises still occur, but they occur earlier, when they can be handled gracefully rather than desperately.

On the other hand, teams who delay seeking help often find themselves reacting to problems rather than planning around them. A formulation must be reinvented mid-way through toxicology. A solid form must be changed during process development. A manufacturing route must be redesigned after scale-up. A regulatory submission must be supplemented with additional studies. Each of these outcomes is avoidable, but only if the warning signs are recognised in time.

The truth is that every molecule will present challenges. Some will be minor; some will be profound. What determines the trajectory of a programme is not the presence of those challenges, but the timing of their discovery. Solid-state experts do not eliminate uncertainty, but they move it upstream, where it is cheaper, easier, and far less disruptive to address.

As drug development becomes more complex, with more insoluble molecules, more sophisticated formulations, more demanding regulatory expectations, etc., the cost of waiting increases. Seeking help is not an admission of difficulty. It is an investment in foresight.

In the chapters that follow, we will turn our attention to the broader perspective: what all of these stories, lessons, and scientific principles teach us about the future. Because solid-state science is not static, it is evolving, integrating more computational insight, more automation, and more strategic thinking. And understanding that evolution is essential for anyone navigating the path ahead.

Chapter 25 — The Future of Solid-State Science: What We Still Don't Know, and Why That Matters

Every scientific field has boundaries, invisible edges where knowledge fades into uncertainty. Solid-state chemistry is no exception. Despite decades of progress, countless case studies, and a steadily expanding analytical toolbox, there remains a simple truth that everyone in this discipline eventually learns: the solid state still holds more questions than answers.

To some, this may sound discouraging. But for those who work closely with molecules, it is what makes the field endlessly compelling. The more we learn about polymorphism, crystallisation, amorphous stability, hydrates, solvates, and molecular packing, the more we uncover complexities that resist tidy explanations. The landscape is vast, and each discovery reveals that it is even larger than we assumed.

What makes the future so interesting is that our uncertainty is not a sign of immaturity but a sign of depth. The solid state is not an orderly domain governed by simple rules, but a dynamic interplay between energy, entropy, kinetics, and environment. Molecules do not line up neatly like soldiers in a parade; they explore configurations, negotiate interactions, and respond to conditions in ways that often escape our initial intuition.

Polymorphism is a prime example. After the ritonavir incident, the industry invested heavily in polymorph screening and modelling. Screening became broader, more systematic, and far more rigorous. Computational tools emerged to predict which packing arrangements might be possible. Automated crystallisation robots allowed teams to explore vast experimental spaces in short timeframes. Yet, even today, no one can guarantee that a polymorph has not been overlooked because the very nature of polymorphism is tied to conditions that may not yet have been tested.

The same uncertainty surrounds amorphous materials. We understand glass transitions, molecular mobility, relaxation, and the thermodynamic drive toward crystallisation. We know how polymers stabilise amorphous dispersions and how humidity accelerates collapse. And still, predicting long-term stability remains challenging. Two formulations that look identical in the first month may diverge after six. A system that appears stable under standard storage may fail under excursions that mimic real-world handling. A polymer that works perfectly for one molecule may be ineffective for another with only subtle differences in structure.

Hydrates and solvates add more nuance. Some form readily, others reluctantly. Some remain stable indefinitely, others transform at the slightest hint of humidity. Some are benign curiosities, others are dominant forms that dictate the entire development path. Despite decades of research, we still cannot reliably predict which molecules will form hydrates, under what conditions, and how stable those hydrates will be.

Even crystallisation, the heart of solid-state science, remains partly mysterious. We understand nucleation, growth, supersaturation, and metastability, but predicting the exact pathway a molecule will take in a reactor, under specific mixing conditions, with real impurities, real solvent gradients, and real scale-up dynamics, is still an art informed by science rather than a deterministic calculation. A molecule can favour one pathway in a 50 mL flask and a completely different one in a 50 L reactor. Modelling helps, but only experiments reveal the truth.

Then there is the question of how solid forms behave in the body. We know that some forms dissolve intact and others convert rapidly. We know that particle size, polymorph stability, and excipient interactions all matter. But predicting in vivo conversion with high confidence remains challenging. Paroxetine's in vivo transformation from anhydrate to hemihydrate is a vivid example, one that the models of its time could not have anticipated. Even today, mimicking such behaviour requires careful experimental design rather than purely computational insight.

The future of solid-state science lies not in eliminating uncertainty, but in managing it more intelligently. Computational methods will grow more sophisticated, but they will not replace experimentation. Machine learning will improve predictions, but it will still need data, high-quality, relevant, diverse data, to learn from. Automation will speed up screening, but interpretation will still require human insight. The scientist's role will not diminish, it will evolve.

What makes this future encouraging rather than overwhelming is that uncertainty is a source of opportunity. Each unknown is a space where innovation can occur. Each surprise is a lesson that refines our understanding. Each failure to predict a behaviour is a reminder of the molecule's complexity, a complexity that, once understood, becomes a competitive advantage rather than a risk.

The molecules we design today are more sophisticated, more rigid, more aromatic, and often less soluble than those of the past. Their solid-state landscapes are richer, more varied, and more sensitive. They challenge us because they demand deeper knowledge, earlier exploration, and more integrated strategies. They force formulation scientists, analytical chemists, crystallisation experts, computational chemists, and process engineers to collaborate more closely than ever.

Perhaps that is the greatest lesson for the future: solid-state science is no longer a specialty working in isolation. It is a crossroads where many disciplines meet. The questions we still cannot answer will be solved not by a single technique or a single perspective, but by the convergence of many. And the more we embrace that convergence, the more predictable, and more innovative, drug development becomes.

As we approach the final chapters of this book, the theme that has emerged again and again is humility. Molecules still have the capacity to surprise us. But they also reward us when we listen carefully, think broadly, and anticipate rather than react. The future will not remove uncertainty, it will simply give us better tools to navigate it.

In the final chapter, we will return to where we began, to the human element. To the experiences, conversations, frustrations, and insights that shaped a career in this field, and to the perspective gained by watching molecules behave well, behave badly, and sometimes behave in ways no one expected.

Chapter 26 — A Career in the Solid State: Reflections, Lessons, and the Road Ahead

When I first stepped into the world of solid-state chemistry, I had no idea how much of drug development lived in these quiet, crystalline details. To the untrained eye, powders look alike, crystals look alike, even tablets look alike. But over the years, I have learned that the smallest differences, i.e. a tilt in molecular packing, a shift in hydration, a slight change in particle size... can alter the entire fate of a drug. And what has stayed with me, more than any single lesson, is that these details are not marginal to development. They *are* development.

Looking back, what strikes me most is not the sophistication of the science, but the human journey behind it. Solid-state work is deeply technical, yes, but it is also profoundly experiential. You learn it not only from textbooks or data, but from the long hours spent trying to understand why a crystallisation that behaved perfectly yesterday produces something unexpectedly different today. You learn it from the moment a routine XRPD pattern suddenly reveals a new peak. From the sinking feeling when a formulation that was meant to be stable crystallises overnight. From the relief when a stubborn molecule finally agrees to crystallise into a stable, reproducible form. And from the quiet, cumulative wisdom that comes from seeing hundreds of molecules move through the same patterns.

Over time, you begin to recognise certain truths. Molecules rarely misbehave without reason. Processes rarely fail without warning. Instability is rarely random. Nature leaves clues, small ones, subtle ones, sometimes frustrating ones… but they are always there. The challenge is learning to notice them early enough, and to interpret them with humility rather than overconfidence.

This book grew out of those moments. Not from a single breakthrough or a dramatic crisis, but from years of working with teams whose only mistake was underestimating how much the solid state matters. I have seen brilliant molecules delayed for reasons that could have been addressed months earlier. I have seen generic manufacturers spend months chasing forms that turned out to be unstable by design. I have seen start-ups lose investor confidence not because their science was wrong, but because their risks were poorly framed. And I have seen experienced teams get blindsided by behaviours nobody expected, behaviours that, in hindsight, were entirely consistent with the molecule's nature.

The purpose of this book was never to create a comprehensive reference. Others have written those with far more depth, and new insights will continue to appear long after this text is complete. Instead, the goal was to capture the essence of what solid-state science means for real-world development: the intersections between solubility, formulation, crystallisation, manufacturing, strategy, and risk. The truths that emerge when a molecule leaves the theoretical world and enters the physical one. The patterns that repeat, no matter the therapeutic area, company size, or technology used.

What I hope resonates is the idea that solid-state science is not something to be feared or postponed. It is something to be embraced. It brings clarity to uncertainty. It turns intuition into evidence. It frees teams from the cycle of reacting to problems and instead allows them to plan with foresight. And it reminds us that a drug is not a structure on a slide, it is a material whose behaviour reflects deep and sometimes surprising physics.

As the field continues to evolve, with computational tools offering new insights and automation accelerating experimentation, the role of the solid-state scientist will change. The questions may become more complex. The datasets will certainly become larger. But the fundamental task will remain the same: to listen to the molecule, to understand its preferences, and to guide it into a form that supports its therapeutic purpose.

My own journey through this field has been shaped not just by successes, but by the challenges that forced me to think differently. Each stubborn molecule, each unexpected polymorph, each puzzling dissolution curve has left an imprint. And each interaction with chemists, formulators, crystallisation engineers, analytical scientists, and regulatory leaders has added another layer to that understanding. No one develops a drug alone. And no one understands a molecule fully until many perspectives come together around it.

If there is one message I hope endures beyond these pages, it is that good decisions in the solid state are not made by rushing toward certainty, but by acknowledging complexity. They are made by asking questions early, by valuing the details others overlook, and by respecting the molecule enough to explore its possibilities before committing to a single path. This is not caution, it is strategy. It is how molecules become medicines.

As the future unfolds, I believe solid-state science will only grow more central to drug development. The molecules in the pipeline demand it. The expectations of regulators demand it. And the needs of patients demand it. But I also believe we are better equipped than ever to rise to that challenge, with tools that reveal the invisible, data that deepen our insight, and a community that recognises the importance of this work.

In the end, what has kept me in this field is not the complexity of the science, but its honesty. Molecules do not lie. They are consistent, even when we do not yet understand them. They behave according to their energy landscapes, not our timelines. And when we listen carefully enough, they tell us everything we need to know.

This book is my attempt to honour that honesty and to share the lessons that have shaped my own understanding of the solid state. If it helps a few molecules reach the clinic more smoothly, or helps a few teams see their challenges more clearly, or inspires someone to look more closely at the crystalline details that matter so much, then it will have achieved its purpose.

The road ahead is full of unknowns, but that is what makes this field so rewarding. The solid state is vast, surprising, and full of possibility. And the more we learn, the more we realise just how much there is left to explore.

Chapter 27 – Carrying the Mindset Forward

Every book ends, but the work it represents does not. The chapters in this final part were written from experience rather than theory, from the patterns that emerge only after years of watching molecules succeed, stumble, or quietly reveal something no one expected. They sit apart from the technical sections not because the science changes, but because perspective does.

If there is one thread running through this book, it is that solid-state science is not a checklist. It is a way of seeing. When you understand how molecules behave as materials, this is, how they pack, transform, dissolve, resist, or cooperate, development becomes less a sequence of obstacles and more a process of discovery. The surprises do not disappear, but they lose their ability to derail. The uncertainties become manageable. The decisions become clearer.

This final reflection is not an invitation to adopt a new doctrine or framework. It is simply a reminder that the most resilient development teams are those that remain curious, that question assumptions early, that recognise when the molecule is trying to tell them something, and that treat the solid state not as a specialist niche but as a foundational part of the drug's identity.

The path from concept to clinic is long, and no two molecules follow it in the same way. But the mindset you bring to that journey, one grounded in observation, humility, and thoughtful anticipation, will shape far more than the choice of a solid form or the outcome of a crystallization. It will shape the confidence with which you move forward, the clarity of the decisions you make, and ultimately the success of the medicine you are trying to bring to life.

If this book leaves anything behind, let it be the reassurance that understanding a molecule is not only possible, but rewarding. And that the deeper that understanding becomes, the more predictable, and more satisfying, the entire process will be.

Acknowledgements

This book would not exist without the many people who have shaped my understanding of solid-state science over the years. I am deeply grateful to the colleagues, mentors, collaborators, and friends who shared their knowledge, challenged my assumptions, and helped me see molecules from perspectives I would never have found alone.

A special thank you goes to my partner, Steve Winter, with whom I had the privilege of founding Solitek. From the very beginning, Steve's vision, energy, and belief in what we could build together have been a constant source of motivation. His commitment to doing things properly, with integrity, curiosity, and a genuine desire to help clients, has shaped not only our company, but also many of the ideas that underpin this book.

To the scientists who work every day in solid-state, formulation, analytical chemistry, process development, and regulatory science: your curiosity and dedication have inspired much of what is written here. Many of the insights in this book were born not in quiet offices, but in shared discussions around laboratory benches, project reviews, scale-up meetings, and long conversations during conference days.

I want to thank the teams, clients, and partners who trusted me with their most difficult problems and allowed me to learn from their molecules as much as from my own experiences. Each project has contributed something to this body of work, and I am grateful for the opportunity to contribute to so many development journeys.

Finally, to my family, my wife Anna and my children, Anna María and Alba, for their patience, encouragement, and belief in the value of doing things properly, even when the work is invisible. Your support has made every page of this book possible.

About the Author

Víctor M. Díaz Pérez is a solid-state and preclinical development specialist with extensive experience supporting small-molecule drug development from discovery through clinical readiness. Over the course of his career, he has worked across medicinal chemistry, crystallisation, solid-state characterisation, formulation development, and CMC strategy, helping both innovative biotech companies and generic manufacturers navigate the complex landscape of the solid state.

Before founding Solitek, Víctor held roles in scientific and development groups where he led or contributed to programmes involving polymorph screening, salt and cocrystal selection, amorphous systems, crystallisation scale-up, developability assessment, and early formulation. His perspective integrates hands-on laboratory experience with the strategic understanding required to guide molecules through real-world development challenges.

At Solitek, Víctor and his team specialise in helping clients understand their molecules deeply, identify risks early, and make decisions that support efficient, predictable drug development. He is a frequent contributor to industry discussions, sharing practical insights on solid forms, particle engineering, formulation challenges, and the intersection between science, business, and regulatory expectations.

When he is not working with molecules or guiding development teams, Víctor enjoys reading, learning, and exploring new ways to make complex scientific ideas accessible to a broader audience. He believes that clarity is one of the most valuable tools in pharmaceutical development and continues to dedicate his time to helping others navigate the solid-state challenges that too often go unnoticed. His work is driven by a simple principle: when we understand a molecule deeply, everything else in development becomes easier, more predictable, and far more rewarding.

References

Amorphous Materials & Solid Dispersions

Baghel, S., Cathcart, H., & O'Reilly, N. J. (2016). Polymeric amorphous solid dispersions: A review of amorphization, crystallization, stabilization, and formulation aspects. *Journal of Pharmacy and Pharmacology*, 68(3), 278–296.

Hancock, B. C., & Zografi, G. (1997). Characteristics and significance of the amorphous state in pharmaceutical systems. *Journal of Pharmaceutical Sciences*, 86(1), 1–12.

Cocrystals, Salts & Supramolecular Chemistry

Childs, S. L., & Hardcastle, K. I. (2007). Co-crystals and their relevance to pharmaceuticals. *Crystal Growth & Design*, 7(6), 1291–1304.

Stahl, P. H., & Wermuth, C. G. (2008). *Handbook of Pharmaceutical Salts: Properties, Selection, and Use.* Wiley-VCH.

U.S. Food and Drug Administration. (2018). *Regulatory Classification of Pharmaceutical Co-Crystals: Guidance for Industry.* FDA.

Computational Tools, CSP & Machine Learning

Aspuru-Guzik, A., Persson, K., & Ceder, G. (2018). Materials science: The AI revolution. *Accounts of Chemical Research*, 51(6), 1285–1292.

Groom, C. R., & Bruno, I. J. (2016). Cocrystal structures in the Cambridge Structural Database. *CrystEngComm*, 18(17), 3485–3496.

Price, S. L. (2014). Predicting crystal structures of organic compounds. *Chemical Society Reviews*, 43(7), 2098–2111.

Crystallisation Science & Process Development

Mullin, J. W. (2001). *Crystallization* (4th ed.). Butterworth-Heinemann.

Myerson, A. S. (Ed.). (2002). *Handbook of Industrial Crystallization* (3rd ed.). Butterworth-Heinemann.

Developability, Solubility & Dissolution

Butler, J. M., & Dressman, J. B. (2010). The developability classification system: Application to formulation and biopharmaceutics decision making. *Journal of Pharmaceutical Sciences*, 99(12), 4940–4954.

Di, L., Kerns, E. H., & Carter, G. T. (2016). *Drug-Like Properties: Concepts, Structure Design and Methods.* Academic Press..

General Solid-State Science & Polymorphism

Bernstein, J. (2002). *Polymorphism in Molecular Crystals.* Oxford University Press.

Brittain, H. G. (Ed.). (2009). *Polymorphism in Pharmaceutical Solids* (2nd ed.). CRC Press.

Newman, A. (2015). Polymorphism in the pharmaceutical industry. *Pharmaceutical Research*, 32(10), 3239–3253.

Nanomilling & Particle Engineering

Merisko-Liversidge, E. M., Liversidge, G. G., & Cooper, E. R. (2003). Nanosizing: A formulation approach for poorly water-soluble compounds. *European Journal of Pharmaceutical Sciences*, 18(2), 113–120.

Moschwitzer, J. P. (2013). Drug nanocrystals in the pharmaceutical industry. *Expert Opinion on Drug Delivery*, 10(3), 409–420.

Preclinical Formulation & Vehicles

European Medicines Agency. (2008). *Guideline on the Non-Clinical Development of Pharmaceutical Products for Human Use*. EMA.

U.S. Food and Drug Administration. (2005). *Nonclinical Studies for the Safety Evaluation of Pharmaceutical Excipients: Guidance for Industry*. FDA.

Regulatory & Quality Guidelines

International Council for Harmonisation. (2000–2017). *ICH Q6A, Q3C, Q3D, Q8, Q9, Q11 Guidelines*. ICH.

U.S. Food and Drug Administration. (1995). *SUPAC-IR: Immediate Release Solid Oral Dosage Forms – Scale-Up and Post-Approval Changes*. FDA.

Case Studies (Ritonavir, Paroxetine, Gatifloxacin)

Bauer, J., Spanton, S., Henry, R., Quick, J., Dziki, W., Porter, W., & Morris, J. (2001). Ritonavir: An extraordinary example of polymorphism affecting drug discovery and development. *Pharmaceutical Research*, 18(6), 859–866.

Chemburkar, S. R., Bauer, J., Morris, J., Henry, R., Spanton, S., Dziki, W., … & Narayanan, B. A. (2000). Dealing with the unexpected appearance of a more stable polymorph during crystallization of ritonavir. *Organic Process Research & Development*, 4(5), 413–417.

SmithKline Beecham Corp. v. Apotex Corp., 403 F.3d 1331 (Fed. Cir. 2005).

SmithKline Beecham plc v. Apotex Europe Ltd. [2004] EWCA Civ

1703.

Byrn, S. R., Xu, W., & Newman, A. W. (2001). Chemical reactivity in solid-state pharmaceuticals: Formulation implications. *Advanced Drug Delivery Reviews*, 48(1), 115–136.

www.ingramcontent.com/pod-product-compliance
Lightning Source LLC
Chambersburg PA
CBHW050109230526
45470CB00004B/1749